WOODEN PLOW PLANES

A Celebration of the Planemakers' Art

Donald Rosebrook
Dennis Fisher

Astragal Press Mendham, New Jersey

The Astragal Press
P.O. Box 239
Mendham, NJ 07945-0239

Cover Design by: Donald Kahn
Photographs by: Dennis Fisher

Library of Congress Control Number 2003101025
ISBN 1-931626-11-1

Printed in Singapore

DEDICATED TO
PATRICIA ANN ROSEBROOK (PAM)
MY WIFE AND BEST FRIEND

Pam contributed greatly to this book through her constant support and her "above and beyond" efforts to scan all of the slides into the computer. Pam as "the Slidesmith" is shown below. She really got into banging the slides with that hammer.

ACKNOWLEDGMENTS

As always I owe thanks to many people, but a few deserve special mention. My wife, Patricia Ann (Pam), assisted me throughout my work on this book. She was my companion on many of the long drives to borrow and return planes. Her work as the "slidesmith," battering down the pegs in plastic slide holders with a tack hammer and a small anvil so they wouldn't jam in the autofeeder during the digitizing step, was critical. She has always been supportive and did everything she could to make my life easier during the writing of this book. I don't think that I ever heard her complain.

Dennis Fisher, my friend and co-author, did nearly all of the photography and supplied the digitizing equipment, while I wrote the text, did the computer photo editing and assembled the book.

Steve Dice reviewed the manuscript and his comments were greatly appreciated. Don Wing provided most of the materials to be used in the section on Greenfield Tool Co., except for the bill head which came from Kendall Bassett. Andy Delans provided all of the material on New York State Prisons, Auburn Tool Co., and on Mr. Embody, Auburn's superintendent. He also provided "work in progress" pictures of a plow plane being made by Bob Baker. Tom Witte provided materials on an early Ohio Tool Co. center-wheel plow plane and pictures of an Ohio Tool Co. plane before it was restored. Bob Baker, planemaker, master restorer, and my friend, provided information on his activities, restored several of my plow planes and allowed us to feature several of the planes he has made.

I am very grateful to the many people that trusted me with their planes so that I could take them to Dennis' home in Roanoke to have them photographed. Among these people were:

Milton Bachelor
Rick Benze
Robert Bills
William Curtis
Andrew Delans
Durward Gehring
Peter Hathaway
James Hill
Mickey Holmes
Mike Humphrey
Frank Kosmerl
Patrick Leach
James Leamy
Robert Nelson
Ronald Pearson
Martyl Pollak
Roy Schaffer
John Sindelar
Roger Springate
John Walkowiak
Tom Witte
Bob Wheeler

Many thanks to the members of the Pacific Northwest Tool Collectors association who put a note in their "Best of the West" meeting announcement, asking members with rare or unusual planes to bring them to the meeting for us to describe and photograph. Among the people who contributed were:

Al Anderson	Steven Dice	Steven Johnson	Kenneth Thomas
Kendall Bassett	David Englund	Frank Manville	Louis Read

Finally, my thanks to Larry Weaver for taking some last minute photographs — the ebony Multiform plow, the ebony Ohio center wheel, and the Robert Baker version of an Israel White patent plow.

Don Rosebrook

When Don asked me if I was willing to do the photography for this book, I agreed, not fully appreciating the extent of that commitment. However, I have to say it has been the most rewarding experience of my tool-collecting "career." What a gift it was to have one of the largest concentrations in the world of very special plow planes covering virtually every available surface in my tool gallery and studio — not once but twice! The best part was the opportunity to get acquainted with these "works of art" through the camera lens. The wealth of knowledge about plow planes that Don articulated during those long photo-shoot sessions made the whole experience even more gratifying. Finally, I want to thank my wife, Glenna, for hosting the "book team" and helping with the editing.

Dennis Fisher

Table of Contents

Foreward

Credit for the concept of this book must go to my wife, Pam, who let me buy almost any plow plane that I wanted as long as I was working. In return, I agreed that I would dispose of those planes, over time, when I retired. When that time came, however, even the thought of parting brought pain. From distress grew the idea that good pictures of the planes would soothe the pain of parting. When I spoke to Dennis Fisher about taking the pictures, I asked if he would like to co-author a book on plow planes. I would scour the country, bring the planes to Roanoke, Va. and he would take the pictures. Beyond that I would prepare the book, since he was still working and I was, at least, trying to retire. In the late winter of 2000, we began as I hauled 80 of my plow planes to Roanoke. In mid-April, I set off on a six week, 11,000 mile odyssey to visit collectors of plow planes and choose pieces from their collections to photograph. By Memorial day, most of the planes were photographed and returned. We photographed some planes in the Pacific Northwest at the 2000 Best of the West Tool Show. Some cleanup work was done in November of 2000, and Pam scanned about 1000 slides into the computer.

Presented with 18 CDs full of digitized slides, I relearned Adobe's Photoshop program and then spent much of 2000 preparing the pictures that you see in this book. In a few rare instances, a plow plane was damaged, yet was possessed of enough beauty or importance to merit inclusion anyway. In these cases, the original picture was "computer enhanced" so that the damage would not be a distraction. In November-December 2001, three more planes were photographed and scanned locally.

There is little text in the book. We preferred, instead, to let the planes speak for themselves. We included close-ups of the details of many of the plow planes in order to emphasize their beauty and the talent of the makers. In addition, we thought that the pictures of the arms, nuts, and moldings would aid researchers and collectors in identifying unsigned planes, of which there seem to be many. We have presented pictures of craftsman-made plows to showcase the ability and imagination of people not in the business of manufacturing planes. The individual presentations often show other details: novel ways of boxing a fence, novel depth stops and novel depth stop locks, and other artful decorations. The book is concluded by four galleries — nuts, arm shoulders, skate noses, and boxing. Although these are intended for the researcher, they demonstrate beauty and fine workmanship as well.

The text concerning the individual planemakers was adapted from information provided in the *Guide to the Makers of American Wooden Planes*, 4th Ed. (Astragal Press). We have used as well the Pollak system of asterisks to denote rarity of makers' planes (for non-collectors, the rarity increases with the number of asterisks — stars to collectors — five denoting less than 10 examples known). We have done this, recognizing that the "rarity" rating applies to all planes by a given maker carrying a certain signature, not just plow planes. It does, however, provide a rough guide to plow plane rarity. If there are less than 10 examples of a planemaker's work known, there are likely only one or two of that maker's plow planes known. It may be that any plow plane with ivory nuts should be rated ***** (most rare) regardless of how many examples of the planemaker's general line are known. But we have not done so and leave that controversial subject to be addressed another day.

We have identified each plane by its owner at the time it was photographed and, if it has subsequently changed hands, by its current owner. For example, "From the collection of Ken Thomas<<<Don Rosebrook" shows that the current owner is Ken Thomas, and that Don Rosebrook was the owner when the plane was photographed. If the plane has changed hands but we don't know the identity of the current owner, the caption would read "From the collection of Don Rosebrook>>>??"

Our greatest regrets are that we were not able to photograph all of the great and interesting planes in the country, and that there was not room in the book for all of those that we did photograph.

We thank all of the people who offered their planes for inclusion in the book.

Introduction

American plow planes evolved dramatically from a primitive device for cutting a flat-bottomed groove, with the grain, at a set distance from an edge, to an elaborate, often beautiful device for cutting that same groove. How is it that such a simple tool, designed to make a cut that, under normal circumstances, no one would ever see, became the most elaborate, most expensive, often gaudiest, woodworking plane that the makers would produce? Was it in the competitiveness of the makers, their artistic skills, the appreciation of the users, or the desire to own something very special? It certainly required disposable income and so, for perhaps the first 100 years of planemaking in America, such tools were not affordable by the working class. But when a class of people appeared that had both the appreciation of beauty and the disposable income that would allow them to buy something special, the planemakers met the challenge.

America's first planemaker never thought of screwarm planes of fancy wood with fancier trim. Francis Nicholson made a slide-arm plow of yellow birch like the one shown at the right. It had an adjustable fence, a depth stop and a way to lock the depth stop. It did the job it was supposed to do.

Plow planes didn't change much in their first 60 or 70 years, as shown by the Gallup plane on the right. They were what are now called slide-arm planes, and the arms (and thus the fence) were locked in place either by thumbscrews bearing on their top surface or by wedges bearing on the sides. They were tools, but some individuality was creeping in. Modest ornamentation was appearing, as was some innovation. A depth stop might be made of bone or perhaps even ivory, or it might have some special design. Some of these are featured in this book.

By 1815, planes, such as that on the right by Alford, looked more polished. The arms were no longer mortised into the fence and the fence now had a leading edge with a molding that made it attractive. The arms might have ferrules or collars on the ends to keep them from splitting. The depth stop was now a cast iron foot fitted with a thumbscrew for adjustment. It may or may not have a side stop or depth stop locking screw. Yes, the plainer planes were still being made and would be for another 80 years, but a lot more choice was now being offered. Planemaking was moving south and west from New England, with New York and Philadelphia also becoming centers of planemaking.

It is believed that screwarm plow planes appeared first in the Philadelphia area somewhere around 1800. Beech was now the wood of choice for the common plane, but the use of boxwood was creeping in and a brass skate-stiffener was becoming more common. Then, around 1820, ivory arm tips appeared, possibly introduced by Collins, however that is by no means proven. Unhandled planes were still the norm.

By the 1840s, many changes had appeared and had become commonplace. Most planemakers made a screw-arm plane and numerous handled planes were found. Ivory trim was being used increasingly. The big manufacturers had appeared, or would shortly, and planemaking became a factory operation. One man would no longer need to make the entire plane. Someone might turn all of the nuts, someone else would make arms and perhaps wedges, and still another person would make the body and the

F. Nicholson ~ 1740-1750

J. Gallup ~ 1790

Alford ~ 1815

Colllins ~ 1820

fence and assemble the plane. The factory might now make its own blades, although not necessarily. Planes were becoming more expensive. By 1855, a simple slide-arm plane would cost $3 or $4, several days' wages for a carpenter.

True innovation began in the 1830's with patents by Israel White and Emanuel Carpenter. Ohio Tool should be credited with the development of the center-wheel plow in the 1850's, in an attempt to make the plane self-regulating (i.e., to have both arms move the same distance as a result of a single adjustment). By the 1870's the Chapin firm developed their "patent adjustable plane." Again a threaded center arm was used to adjust both arms simultaneously but, unlike the others, none of the arms now extended through the body, so the user could work very close to any obstruction on the right side of the plane (his left side).

By the 1900's there were very few wood plow plane makers left in the United States. Stripped of their fancy decoration, the planes might look much as they had for the previous 50 years. The rosewood plow by Sandusky Tool Company is an example of a fine screwarm plow from that era.

Throughout this time period, some people made their own planes. This was accomplished with varying degrees of skill and with the use of decorative materials similar to those used by the manufacturers. The craftsman sometimes produced his own version of patented planes and often produced a plane of fine functionality, good imagination, and surprising beauty. A few talented people still produce wooden plow planes today. Some of them are meant to be used, while others are meant to show the skill and artistry of their makers. Such a plane is shown below.

This brief look at plow plane making in the United States cannot hope to cover all of the aspects of how they were constructed, who made them, and how planemaking moved across the country. Thus we rely on the photographs in the book to convey the evolution, innovation, beauty, artistry, and ingenuity of the makers.

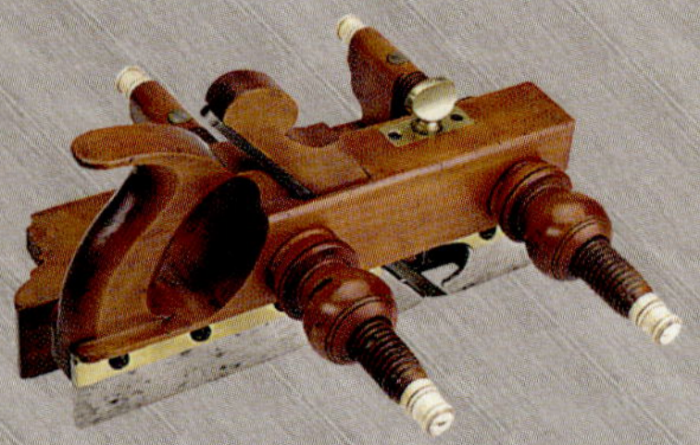

Casey, Kitchell & Co. ~ 1850

Ohio Tool Co. ~ 1870

Sandusky Tool Co. ~ 1900

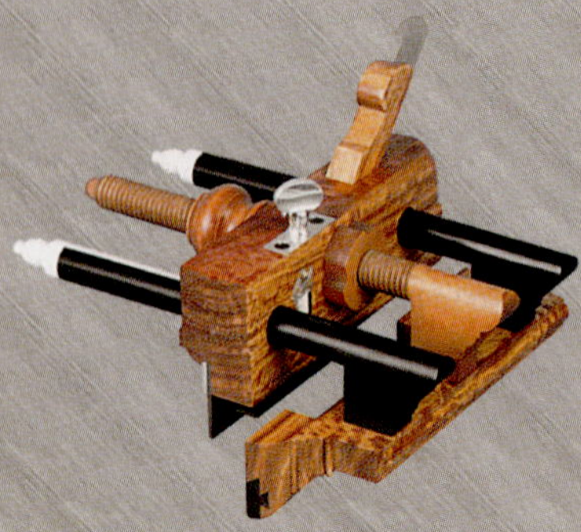

R. Baker 2001

1 New England Planemakers

New England, specifically Massachusetts, is recognized as being the cradle of American planemaking. Two things stand out about New England plow planes. The first is the "Yankee Plow," an early style of plow plane found only in New England. The second is the particularly fine quality of the products of many of the manufacturers.

"Yankee Plows" were characterized by a fence the same length as the body, arms mortised into the fence, and usually (but not always) wide flat chamfers on the edges of the body. The earliest planes were made of yellow birch, which gradually gave way to beech.

Nineteenth century New England planes showed the influences generated within smaller regions of that section of the country. For example, the planes of Rhode Island were made primarily in Providence and show great similarity in many characteristics, such as the turned up skate nose. The planes of southern Massachusetts show the influence of the Lamb family. The planes of Boston show similarities, as do the planes of central Connecticut. This should not be construed to mean that the planes look identical, but instead that there are some features among the group that are similar.

High quality planes were a hallmark of the Lambs and their various companies, as well as of the Greenfield Tool Company, DeForest, Chapin and several others. The photographs will demonstrate that fact. Greenfield Tool Company is profiled in this section and many of their fine planes are shown.

Chapin is noteworthy not only for nearly 100 years of planemaking but also for the planes utilizing the Chapin-Rust patent. This patent allowed them to produce planes with no obstructions on the left side, allowing operation very close to any obstruction from the piece being planed. Three such models were produced and sold for many years, and a "modern" bridle plow was made as a probable extension of the third model of that plane.

Thomas Worrall of Boston patented a feature which allowed interchange of handles and plane bottoms, and although Ohio Tool Company did make some planes utilizing this patent, it never became popular among other planemakers.

Ivory was used liberally on many New England planes, but, strangely, Chapin is not known to have produced a plow plane with ivory. Greenfield Tool Company, DeForest, Sandford, and perhaps others used ivory liberally to decorate their plow planes, even to the extent of making ivory nuts. An ivory wedge is known on a Lamb plane. There was frequent use of nickel plating of brass pieces on the high end planes in the product lines of several planemakers. This was called "silver mounting." Only a very few special planes actually used silver for plating.

The Greenfield Tool Company

It might be said that the Greenfield Tool Company started in a blaze of glory and ended in the same fashion after 35 years in business. The predecessor companies were Parker, Hubbard & Co. in 1842, and the Conway Tool Company in 1850. Fire destroyed the Conway Tool Co. factory on July 6, 1851. The stockholders were Alonzo Hubbard, Horace Hubbard, and Daniel Rice. On July 28, 1851, they decided to rebuild in Greenfield, Mass. and tool manufacturing was begun sometime in 1852,[1, 2] after which they became one of the largest plane manufacturers in America. The company apparently had a rapid rise toward success and high production, but a series of financial difficulties lead to the loss of the company through insolvency in 1883. A disastrous fire on March 24, 1887 destroyed the entire plant.

South View of the Greenfield Tool Co.'s Works *(Courtesy of Donald & Anne Wing)*

The Greenfield Tool Co. factory was said to be 200 feet by 25 feet and, as shown in the above view, most of it was two stories tall, resulting in a large amount of floor space. Some of these buildings remain in place today (the brick parts), although much modified.

Although it does not concern us here, the company made many products other than planes. Many of them are shown on the bill head presented opposite, which advertises a wide mix, from ox shoes to ivory handled cutlery to bread cutters. Also note that the Chambers Street address in New York City was crossed out by late 1884. However, joiners planes apparently were still the primary product.

The company offered for sale a large variety of plow planes ranging from "4th rate slide-arm plows" to a handled silver mounted, all ivory plow. While it seems doubtful that the ivory plow was ever produced, perhaps one will appear somewhere, someday. The company's catalog of 1854 offered 58 varieties or styles of plow planes; the catalog of 1872 offered the same planes minus the ivory plow. The

---

[1] Roberts, Kenneth, *Wooden Planes in 19th Century America*, Vol. 1, 2nd Ed. Ken Roberts Publishing Co., Fitzwilliam, N.H. 1982, pp 27-28.

[2] Wing, Donald & Anne, "The Mechanick's Workbench," Catalog No. 10., Marion, Mass. August 1980, p 3, 10, 11.

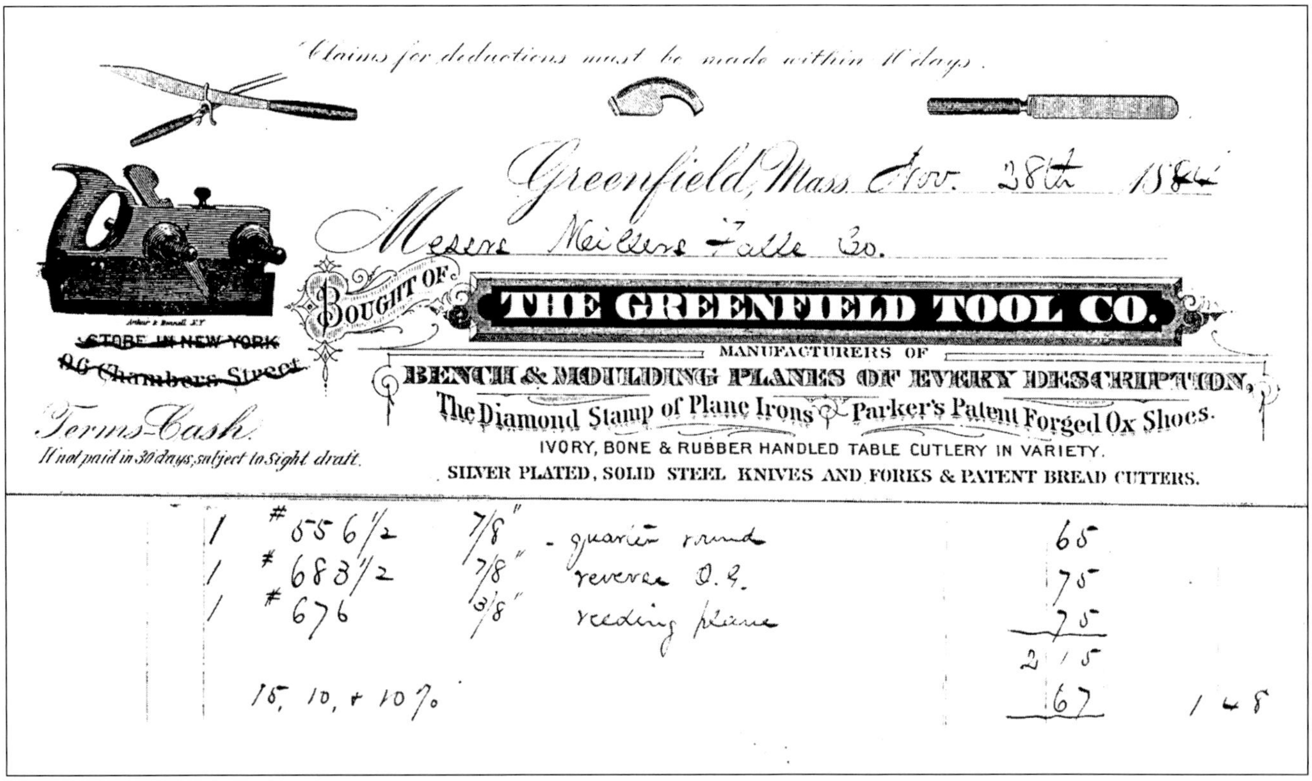

Claims for deductions must be made within 10 days.

Greenfield, Mass. Nov. 28th 1884

Messrs Millers Falls Co.

Bought of THE GREENFIELD TOOL CO.

MANUFACTURERS OF

BENCH & MOULDING PLANES OF EVERY DESCRIPTION,

The Diamond Stamp of Plane Irons & Parker's Patent Forged Ox Shoes.

IVORY, BONE & RUBBER HANDLED TABLE CUTLERY IN VARIETY.

SILVER PLATED, SOLID STEEL KNIVES AND FORKS & PATENT BREAD CUTTERS.

STORE IN NEW YORK 96 Chambers Street

Terms Cash.

If not paid in 30 days, subject to Sight draft.

1	#556½	7/8"	quarter round	65
1	#683½	7/8"	reverse O. G.	75
1	#676	3/8"	reeding plane	75
				215
	15, 10, + 10%			67
				148

Invoice from the Greenfield Tool Co. ***(Courtesy of Kendall Bassett)***

model numbers had, however, changed and the 1872 model numbers were all less by 19. The high end plow was a handled ebony plow — ivory trimmed, silver mounted and boxed — that listed for $18.30 in 1854 and $22.80 in 1872. The ivory plow was listed at $75 in 1854. In general, the prices in the catalogs seem competitive with those of other planemakers.

The principal difference between Greenfield Tool Co. and everyone else was the number of high end (i.e., rosewood or ebony with silver mounting and or ivory) plows offered by Greenfield. They offered 26 models of rosewood or ebony plows, 17 of which contained ivory or ivory and silver mounting. The planes pictured following this feature depict many of those model numbers. In addition, the pictured planes contain features not listed in the catalogs (but found in the ledger), such as boxwood plows with ivory trim, ebony plows with ivory trim and ivory tips. This can be interpreted in several ways — perhaps these plows were special order, or there were models listed in catalogs produced between 1854 and 1872, or perhaps they were produced before the 1854 or after the 1872 catalogs. The ebony plow shown on pages 18 and 19 is marked as model No. 553, which would be the top of the line in the 1872 catalog. Because of the large variety of ivory containing planes and because the company made planes for others to sell (such as D. R. Barton, see page 9), opportunities for interchange of parts were great. Regardless of how they came about, the individual styles of plows containing ivory are rare.

A distinctive feature of Greenfield Tool Co. plows is the use of sandwich nuts (the catalog calls them ivory banded nuts). Here again the catalog does not list any ivory banded boxwood nuts. The picture on the next page, however, is ample evidence of the fact that such nuts were produced. The general shape of all of these nuts is similar, although height and width vary widely. Only the 4th and 5th nuts from the left have truly similar dimensions. This implies that the nuts were produced using scraps of ivory and, rather than have wastage, the dimensions of the nuts were allowed to change.

Don and Anne Wing have graciously shared their material on the Greenfield Tool Company in the form of a company ledger covering a period of 20 months from April 1, 1853 to November 30, 1854. Their discussion of the material in this ledger can be found in "The Mechanic's Workbench" (footnote 2 on page 4) which contains a more far-reaching and complete discussion of the company's activities during that period. The Wings' analysis of the information is recommended reading for anyone interested in the history of wooden planemaking. Here we shall be concerned only with the information as it applies to plow planes.

Ivory Banded (Sandwich) Nuts

In the ledger provided by the Wings, only two ivory trimmed plows were noted in the 20 months covered by the ledger. Both were handled ebony plows with ivory trim. One had silver mounting, the other did not. They were both made by Lucius Graham during April 1853; he was paid $1.75 each, the same price as that for any other ebony or rosewood plow that he made. This implies that he did not make the nuts, but that they were provided to him. He would have received $1 for a first rate beech screw-arm plow and $1.34 for a first rate handled beech screw-arm plow. Unhandled boxwood screw-arm plows earned a planemaker $1.33 and handled plows earned him $1.70. Ivory tips added $.04, so the unhandled boxwood plows with ivory tips earned their maker, Levi Gunn, $1.37 each. Producing an unhandled rosewood plow would earn the planemaker $1.35, his payment rising to $1.38 if the plane had sidescrews (called depth stop locks elsewhere in this book). Boxing (boxwood inserts along the face of the fence) would earn $.01 more for the planemaker, and the addition of plating (brass stiffener added along the top of the skate) would add an additional $.04.

The factory workers apparently were making about 16 cents per hour when they were paid on a time basis. That is about $10/week for something in the neighborhood of an 11 hour day, six days per week (about $43/month). Planemakers could make a better wage doing piece work; for example, in September of 1853, Levi Gunn was paid $63.47 for 78 plows of various types. On the other hand, the planemaker could also make a lot less; for example, in November of 1853 Levi Gunn was paid only $19.12 for 19 plows of various types.

In 1854, plow planes prices ranged from as little as $2.85 for the cheapest plow to $4.25 for a first rate unhandled beech plow — boxing or plating would add $.30 to the cost. If you wanted a handled 1st rate beech plow, the cost was $5.75, with the same $.30 extra for boxing or plating. Unhandled boxwood and rosewood plows sold for $6.50, while ebony sold for $6.75. Handled boxwood and rosewood plows sold for $8.20, while an ebony plow sold for $9.50, or about a week's wages for a skilled worker.

Ivory trim (taken to mean ivory nuts) added approximately $6 to the selling price of a rosewood or ebony plow. The material cost of ivory appears to be the dominant factor, since one could have ivory banded nuts added for about $2.50, despite the additional labor required. So-called silver mounting, which was a plating operation, added approximately $2.50 to the cost, a seemingly excessive amount. If you wanted a fancier plow, you could count on roughly doubling the price to go from beech to ebony and roughly doubling that again if you wanted your ebony decorated with ivory and silver mounting. Thus, one paid a high price for vanity features.

Much of the Wing's analysis in their article referred to above concerned the way that the Greenfield Tool Company did business. They found that in the first 12 months of the time period covered by the ledger, all of the plows were made by the same five men — Lucius Graham, Charles Amidon, George H. Phillips, Thaxter P. Grout and Levi Gunn. During that time these men made 1,843 plows of various types and styles, and not all of them made plows every month. The following table is an analysis of the plow plane production, taken directly from the Wings' article.

	Beech Screw-Arm	Beech Slide Arm	Boxwood	Rosewood	Ebony
Handled	309	0	102	67	22
Unhandled	701	217	254	115	56
Total	1010	217	356	182	78

The overall plane production of all kinds of planes for the first 12 months would seem to be about 130,000 based on the two months total production presented by the Wings. Thus, plow planes may have accounted for only about 1.4% of the total production. Using the production numbers in the table above, one can estimate that ebony planes comprised 0.06% of the production, and handled ebony planes comprised 0.017% of the production. Given that only two of the ebony and rosewood planes utilized any ivory, this means that they comprised 0.0015% of all production.

There is no legitimate way to extend this information to cover the entire lifetime of Greenfield Tool Company, but it would seem that only a very few planes were trimmed in any way with ivory. Given that some of them had "silver mounting," ivory banded nuts, solid ivory nuts, or boxing, one can arrive at 8 versions of handled ebony and ivory planes and 10 versions of rosewood and ivory planes. The logical conclusion is that demand for — and thus, production of — a plow plane that could cost up to two weeks' wages for a skilled workman was extremely small. This limited production leads to scarcity and should explain why ivory trimmed plows are so prized by collectors today.

There were very few planemaking operations as large as Greenfield Tool Company. Perhaps only the New York Prison operations, the Ohio and Sandusky Tool Companies and H. Chapin produced planes on the same scale. Of these, H. Chapin is not known to have produced planes with ivory trim of any kind. The ivory trimmed planes of the others are about as rare today as those of Greenfield Tool Co. In the case of small planemaking concerns, it is assumed that they made no more than one or two such planes during their entire operating period, and many made none. The planes presented in this book by such small operations as Wm. Goldsmith, Wm. Epperson, R. J. MacCubbin, and A. McBride, may well have been the only plow planes with ivory trim made by these makers.

Greenfield Tool Co.

Handled boxwood screw-arm plow No. 532.

(From the collection of Don Rosebrook)

Greenfield Tool Co.

Handled beech screw-arm plow No. 526, with boxwood wedge, arms and nuts. Overstamped D.R. Barton, but the Greenfield mark is still visible.

(From the collection of Steven Dice <<< Don Rosebrook)

Greenfield Tool Co.

Handled boxwood screw-arm plow, No. 516, with rosewood nuts.

(From the collection of Don Rosebrook)

At first glance, the nuts would not seem to belong to a Greenfield plane, but other planes are known to the authors with the same style of nuts and arm tips. Also, the model number 516 is a number that cannot be correlated with models in known Greenfield catalogs from 1854 and 1872, although here again, other Greenfield planes with non-conforming model numbers are known and several are shown in this book. The arm tips can be traced back to Parker & Hubbard, a predecessor company.

Greenfield Tool Co.

Unhandled rosewood screw-arm plow, No. 539, with boxwood arms and boxwood flat nuts, ivory tips, and sandwich nuts of ivory and boxwood.

(From the collection of John Sindelar <<< Don Rosebrook)

The model number corresponds to a plane in the 1854 catalog.

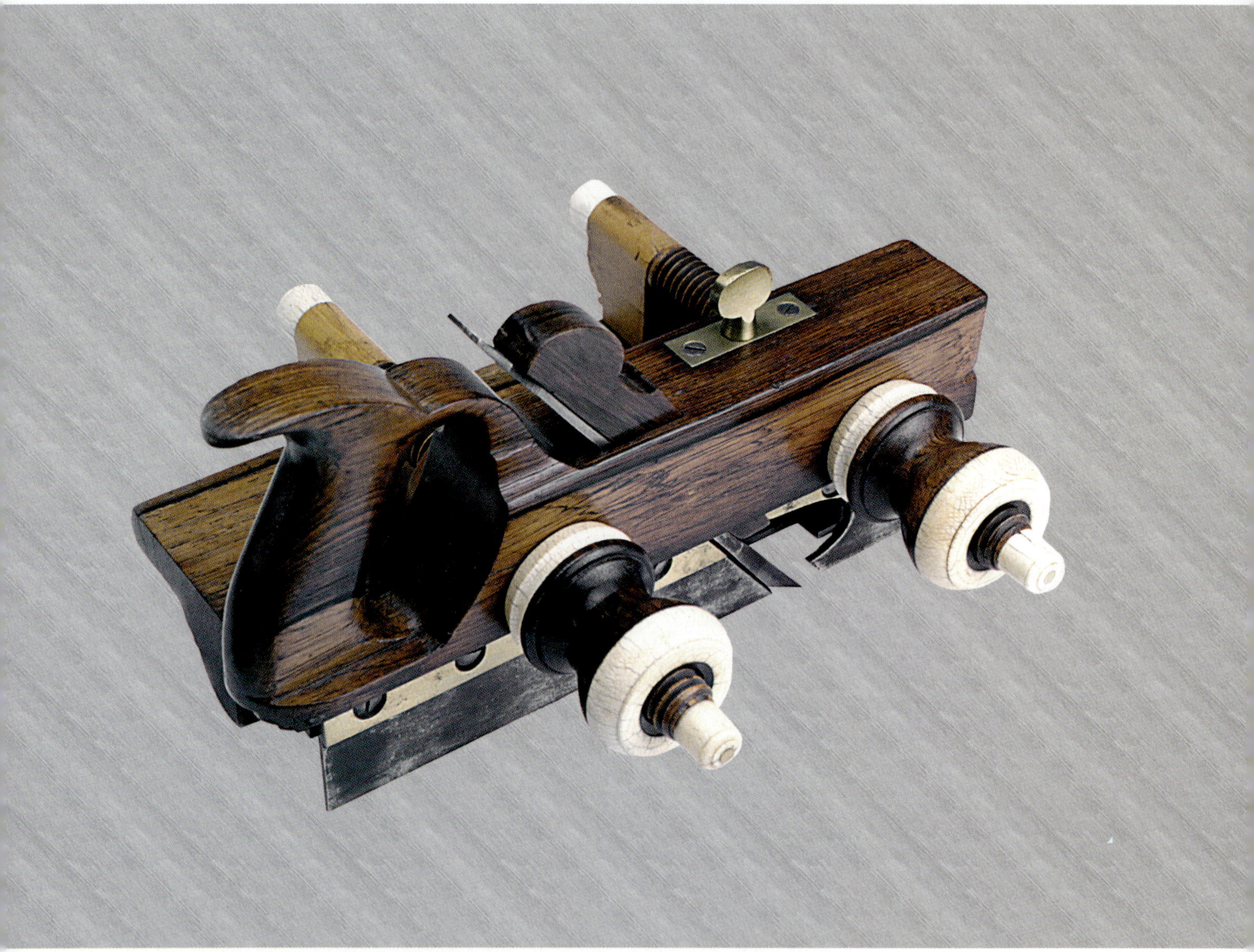

Greenfield Tool Co.

Handled rosewood screw-arm plow, No. 537, with boxwood arms and flat nuts and ivory tips. The large nuts are a sandwich of ivory and rosewood between ivory pieces. The plane is marked D. R. Barton.

(From the collection of Dennis Fisher <<< Don Rosebrook)

There is no doubt that the plane was made by Greenfield. See the beech Greenfield No. 526, page 9, for further comment.

Sandwich nuts of ivory and boxwood, rosewood, or ebony are a feature of many of the Greenfield higher quality planes. Although a few other Massachusetts planemakers made a sandwich nut, the construction of these nuts is different, in that Greenfield used a wooden center with a flange on each end, onto which was slipped an ivory ring of distinctive shape.

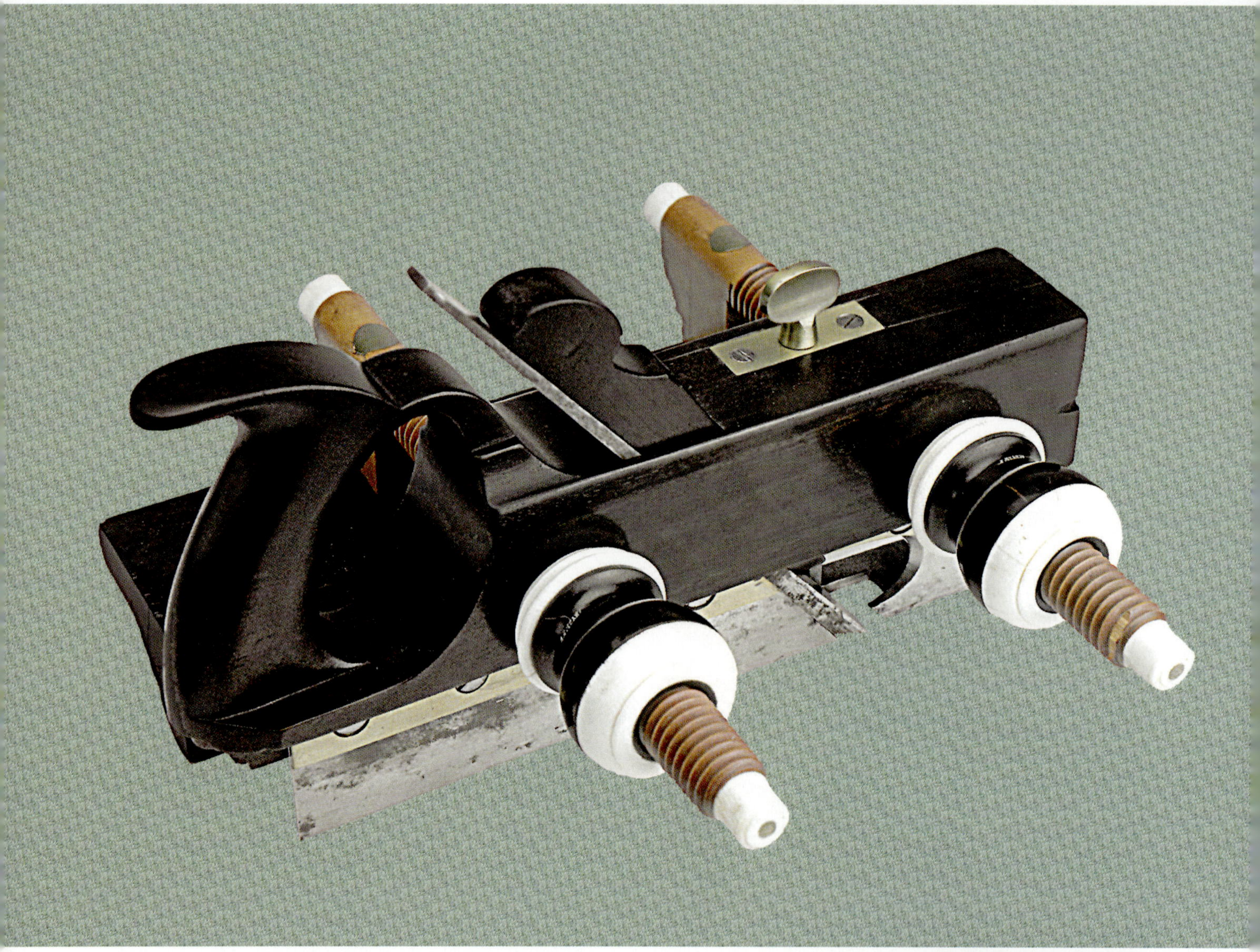

Greenfield Tool Co.

Unsigned handled ebony screw-arm plow with boxwood arms and ivory tips. Nuts are a sandwich of ebony between ivory pieces.

(From the collection of Don Rosebrook)

Among the features that make Greenfield plows so striking are the full width platform for the base of the handle, and the fence that extends to the rear of the handle base. These two features combine to make the Greenfield plows look much more massive than most other contemporary plows. Along with the distinctive sandwich nuts, these features combine to identify the maker of this plane.

Greenfield Tool Co.

Handled rosewood screw-arm plow, No. 542, with boxwood arms and ivory nuts and tips.

(From the collection of Don Rosebrook)

The large ivory nuts on this plane were turned so thin in the center that they have become translucent.

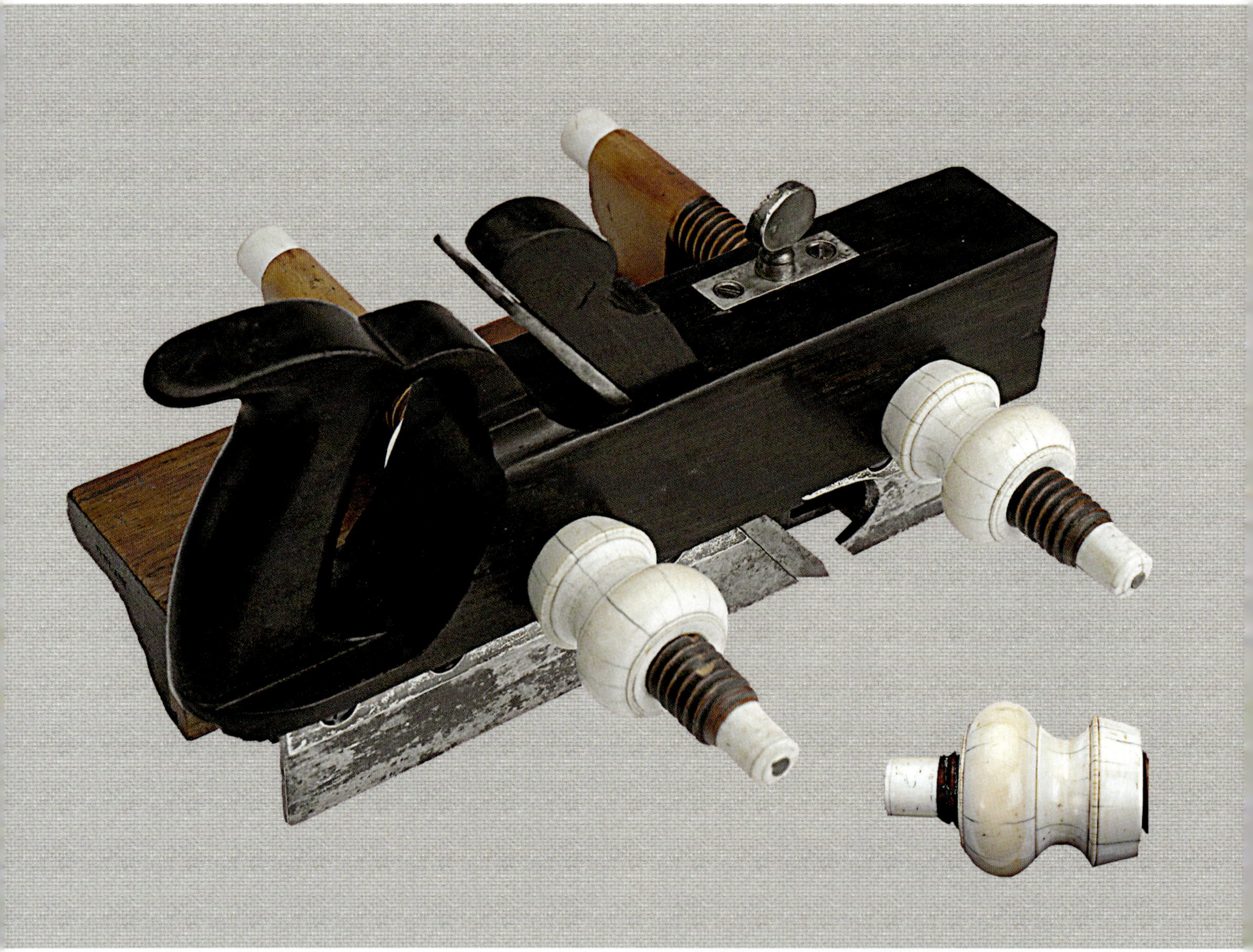

Greenfield Tool Co.

Handled ebony screw-arm plow No. 553 with a boxed rosewood fence, boxwood arms, ivory tips and nuts, and silver mounting.

(From the collection of Don Rosebrook)

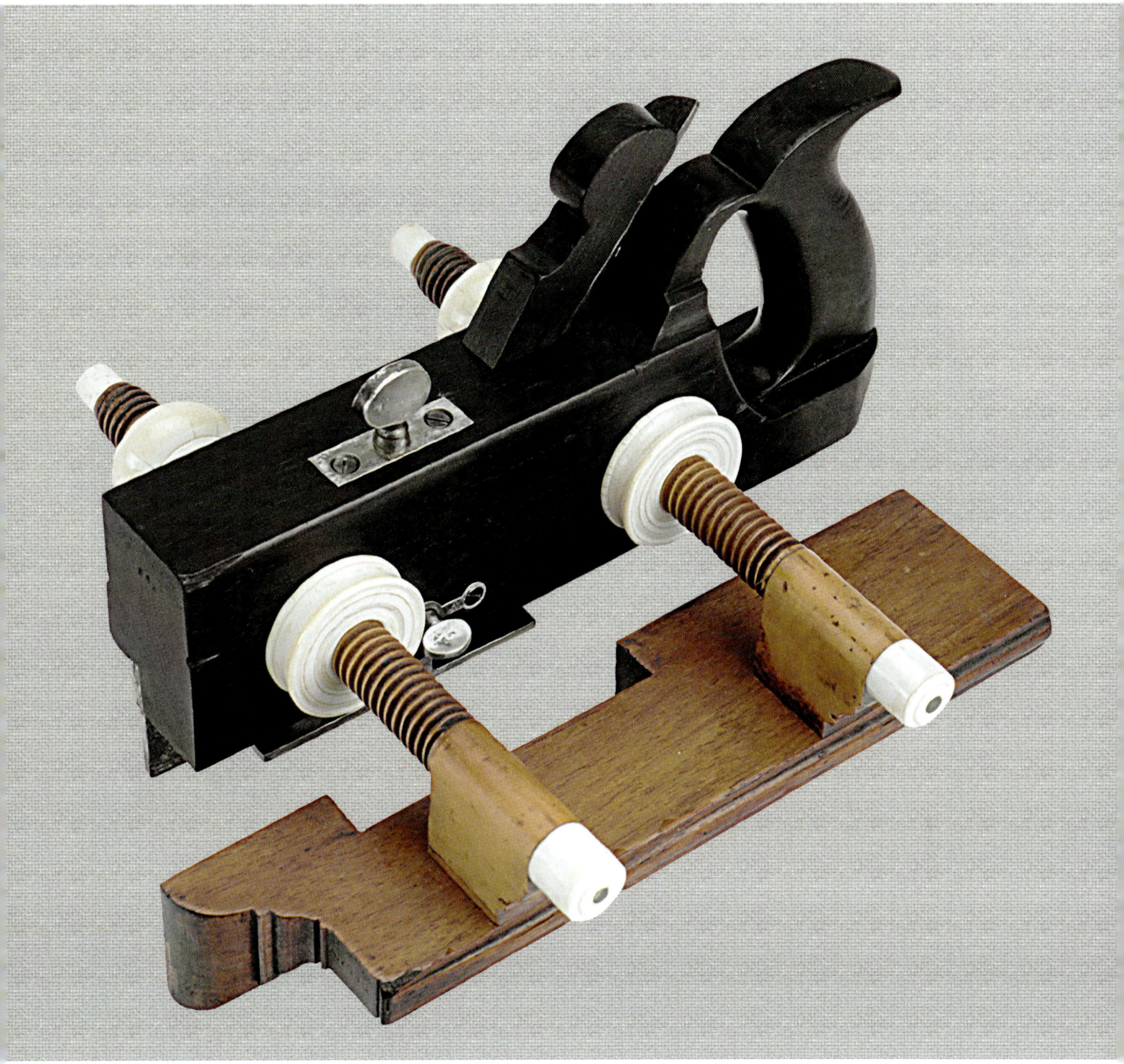

The silver or nickel plating on this plane has worn until it has become quite ragged, although it is obvious that it originally was nickel plated. The rosewood fence could well be original to the plane.

Greenfield Tool Co.

Handled ebony screw-arm plow, No. 522, with boxwood arms and ivory nuts.
(From the collection of Don Rosebrook)

This is another of the series of planes with model numbers that do not correlate with the two known catalogs. Again, the arm tips provide a clue to membership in the group of non-correlating model numbers, as does the shape of the large ivory nuts.

Greenfield Tool Co.

Handled boxwood screw-arm plow, No. 532, with ivory nuts.

(From the collection of Don Rosebrook)

Model 532 is listed in the 1872 catalog as a handled boxwood screw-arm plow. There was never a listing of a boxwood plow with ivory. The nuts, however, do have a superficial resemblance to later Greenfield nuts. Nuts containing ivory seem to have been turned from scraps of ivory; the shape depended on the size and shape of the original ivory piece. See the gallery of ivory sandwich nuts.

The nuts have a strong resemblance to those of Kellogg. (Which see.)

Greenfield Tool Co.

Handled rosewood screw-arm plow, No. 536, boxwood arms and flat nuts, rosewood and ivory sandwich nuts, and boxed fence.

(From the collection of Steven Dice <<< Don Rosebrook)

Corresponds to the number in the 1872 catalog.

Greenfield Tool Co.

Handled rosewood screw-arm plow, No. 551, boxwood arms and flat nuts, and boxwood and ivory sandwich nuts.

(From the collection of John Sindelar <<< Don Rosebrook)

Corresponds to the number in the 1854 catalog, except that one expects rosewood sandwich nuts.

Greenfield Tool Co.

Handled rosewood screw-arm plow, No. 536, boxwood arms and flat nuts, rosewood and ivory sandwich nuts.

(From the collection of John Sindelar <<< Don Rosebrook)

Corresponds to the number in the 1872 catalog, except that the fence is not boxed and thus should be No. 535.

Gladwin & Appleton

Handled boxwood screw-arm plane. *

(From the collection of Don Rosebrook)

Gladwin & Appleton was a partnership of Porter Gladwin and Thomas Appleton that made planes in Chelsea, Massachusetts from 1873-1877.

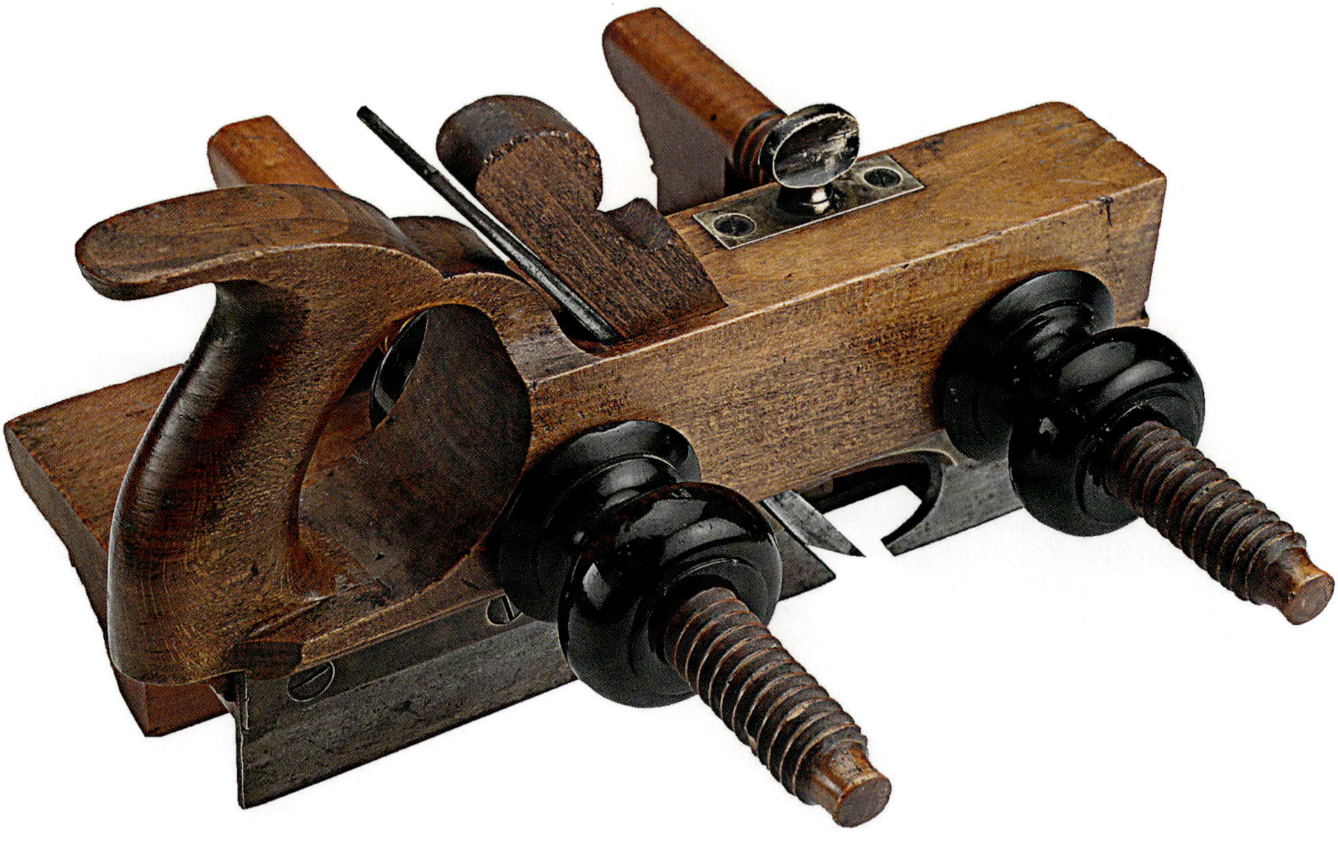

Parker & Hubbard

Handled beech screw-arm plow plane with rosewood nuts and boxwood arms.

(From the collection of Steve Dice)

Alonzo Parker and Horace Hubbard were tool manufacturers (wooden planes were among their tools) in Conway, Massachusetts from 1842-1849. In 1850, they founded the Conway Tool Co. (another plane maker) and their facility burned in July of 1851. Afterwards, Parker moved to Greenfield Massachusetts and founded the Greenfield Tool Co.

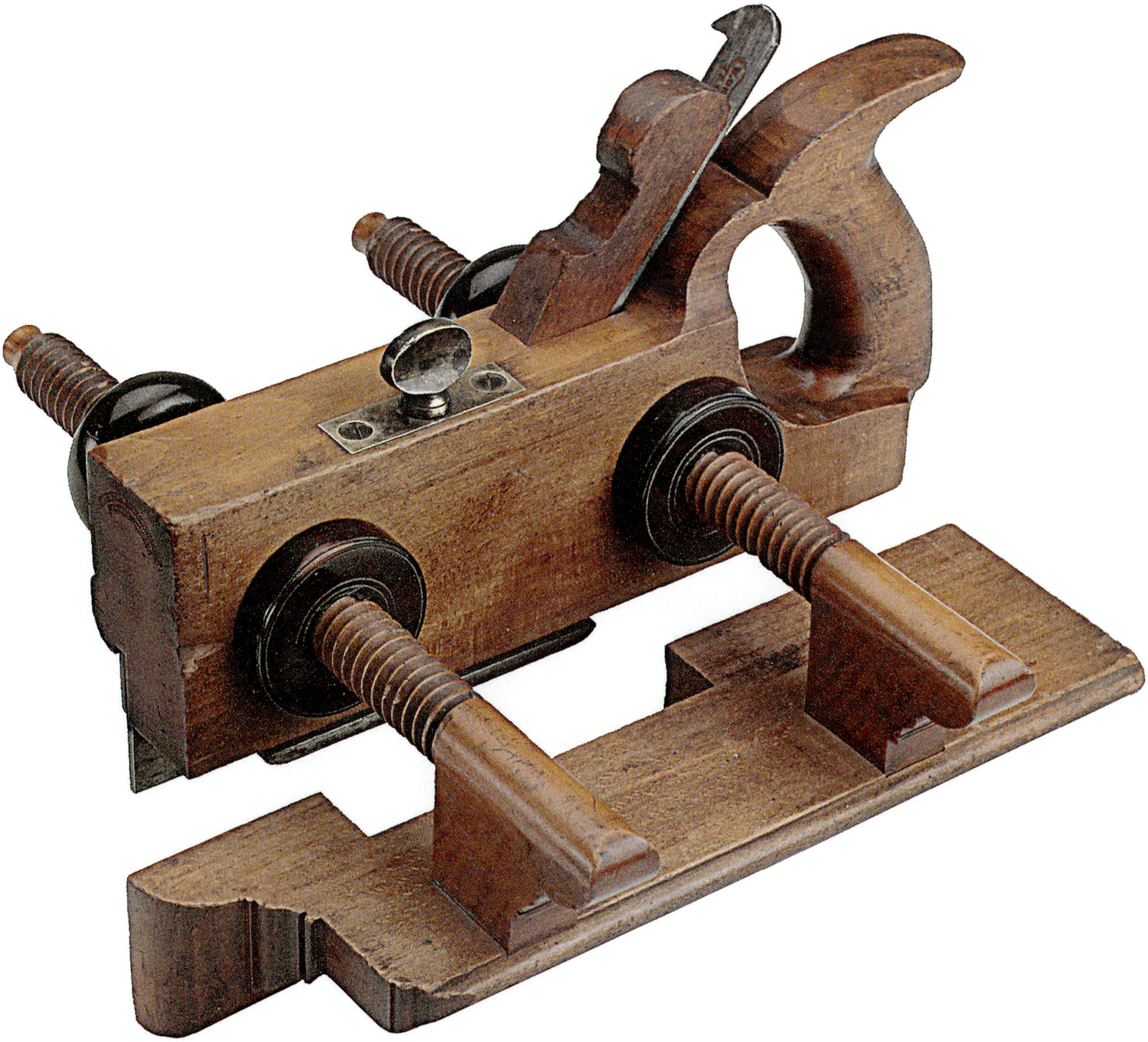

One look at this plane and one of the mysteries of the style of the Greenfield planes is solved. The shape of the nuts and arm tips of early Greenfield planes has its foundation in Alonzo Parker's experience. One might expect that the model numbers on some of the Greenfield plows result from a much shorter list of available models. That would well be the case for a company just starting production.

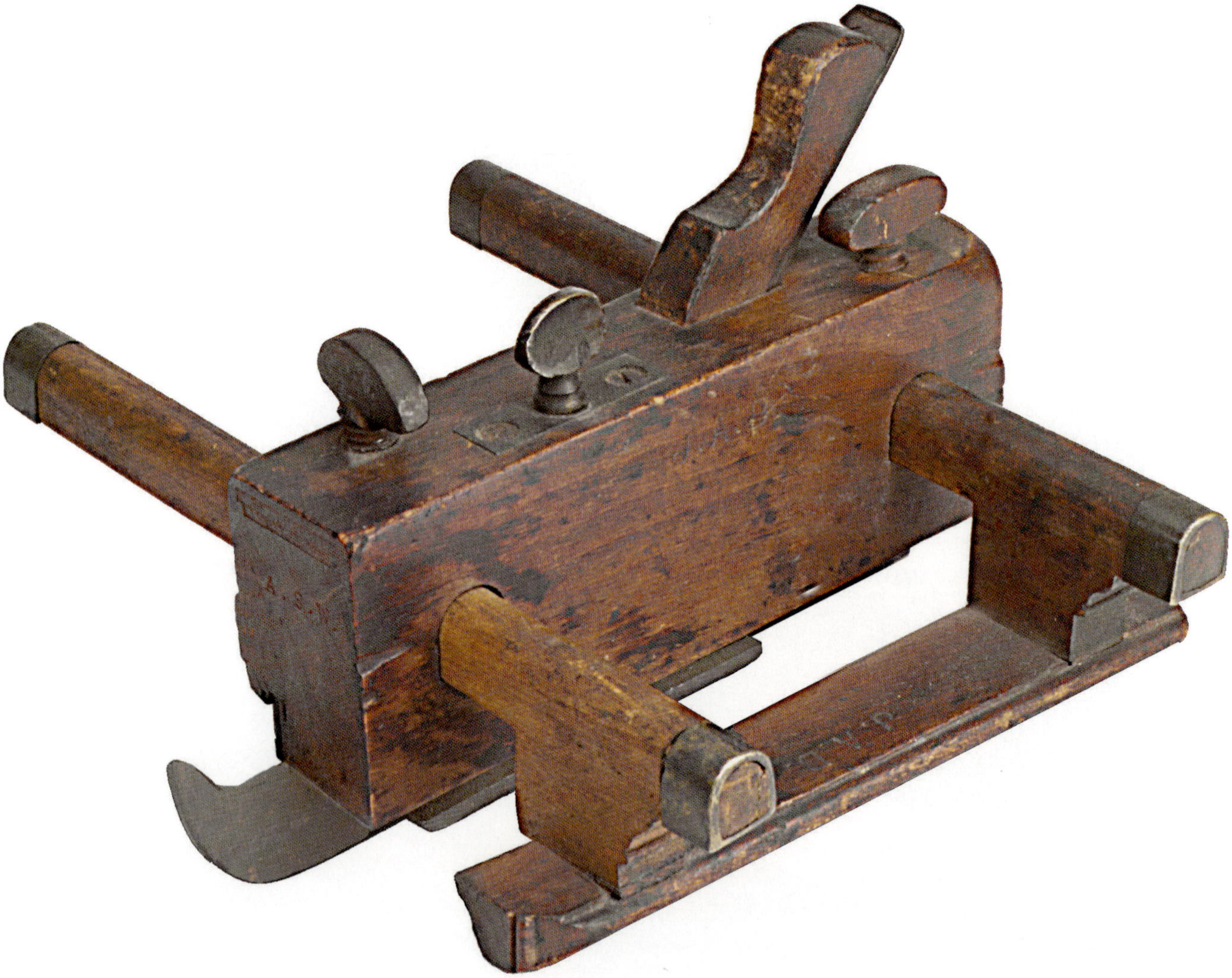

James Kellogg

Unhandled beech screw-lock slide-arm plow. Skate nosed.

(From the collection of Jim Hill)

The nose on the skate of this plow is unusual. Not only are skates with turned up noses unusual, but this one has a crude look to it and does not flow smoothly from the nose of the plane.

James Kellogg

Unhandled rosewood screw-arm plow with boxwood arms and nuts.

(From the collection of Don Rosebrook)

James Kellogg made planes in Amherst, Massachusetts from 1840-1864 under his own imprint. He also worked in various partnerships before and after this period. His operation was apparently quite large. Nuts made in the Kellogg pattern have been noted on Greenfield Tool Co. plows including one of ivory shown earlier. Roger Smith references an apparent connection between the two companies through a man named R. B. Rice who worked for both companies and others.

James Kellogg

Unhandled beech screw-lock slide arm plow, beech depth stop and beech thumb screw stop lock.
"A" mark **

(From the collection of Peter Hathaway)

Note the narrow mortise in the fence to accommodate the depth stop. This feature is very unusual but that is all the room that is needed for the narrow wooden depth stop. All planes with a wooden depth stop could be made this way because the stop is much narrower than the standard cast iron stop.

Copeland & Co.

Handled mahogany screw-arm plow with rosewood fence and boxwood wedge, arms and nuts, and depth stop lock.

(From the collection of Don Rosebrook)

Copeland & Co operated from 1855-1866 in Huntington, Massachusetts It was the successor to M. Copeland & Co, a firm operated by the Copeland brothers — Melvin, Daniel and Alfred. Since Daniel died in 1854 and Alfred in 1858, it would appear that Copeland & Co. was a continuation of the earlier partnerships. Copeland is noteworthy because it made bench planes using a Worrall patent. Worrall's patent plows are presented as Multiform Molding Co. plows on pages 53-55.

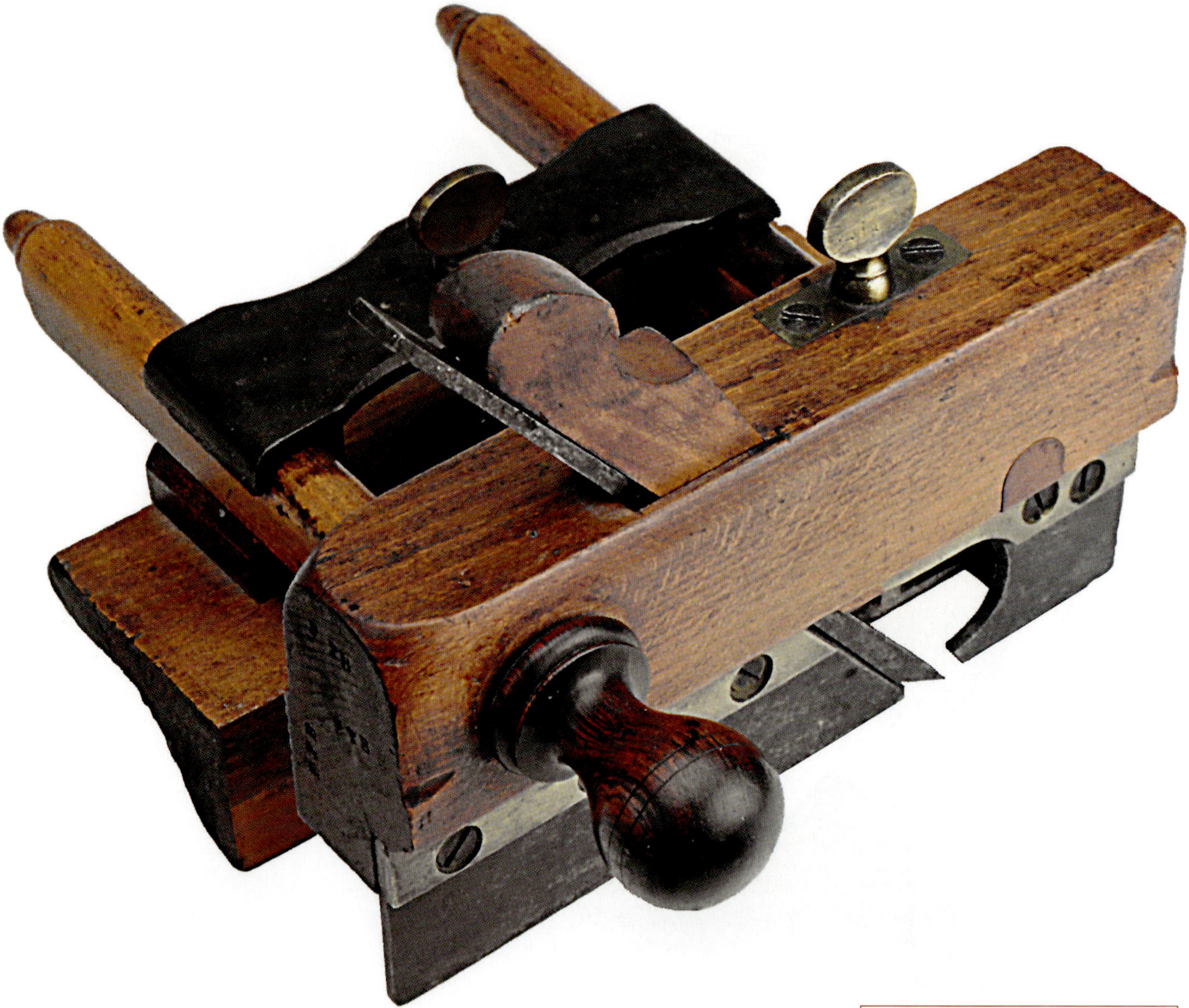

A. Cumings

Unhandled beech bridle plow with cast iron bridle, rosewood side-handle, and boxed fence. *

(From the collection of Don Rosebrook)

A. Cumings operated in various partnerships in New York and New England, but under his own name in Boston during 1848 - 1854. Pollak presents evidence of interactions between Cumings and the Greenfield Tool Co.

This plane, called a "bridle plow," is a very rare style in American plane making. The fence is captured and locked to the arms by an iron crosspiece called a bridle. In theory, this allows the fence to be adjusted by loosening a single thumb screw. Without deeply contoured arms and bridle, it would have been difficult to maintain the fence parallel to the skate during adjustment. This style was popularized by Mathieson and Co. in Scotland many years later. In the Cumings plane, the arms pass completely through the body, and their ends can be seen in the view on the opposite page.

A. Cumings

Handled boxwood screw-arm plow with rosewood fence, boxwood wedge and nuts, and arms made of satinwood or mahogany. *

(From the collection of Milton Bacheller)

The arms are mortised into the fence and are made of a wood of unknown identity. Several woods are known to have been used for arms by various planemakers. For example, Denison is known to have used dogwood for arms while other planemakers are known to have used West Indian satinwood.

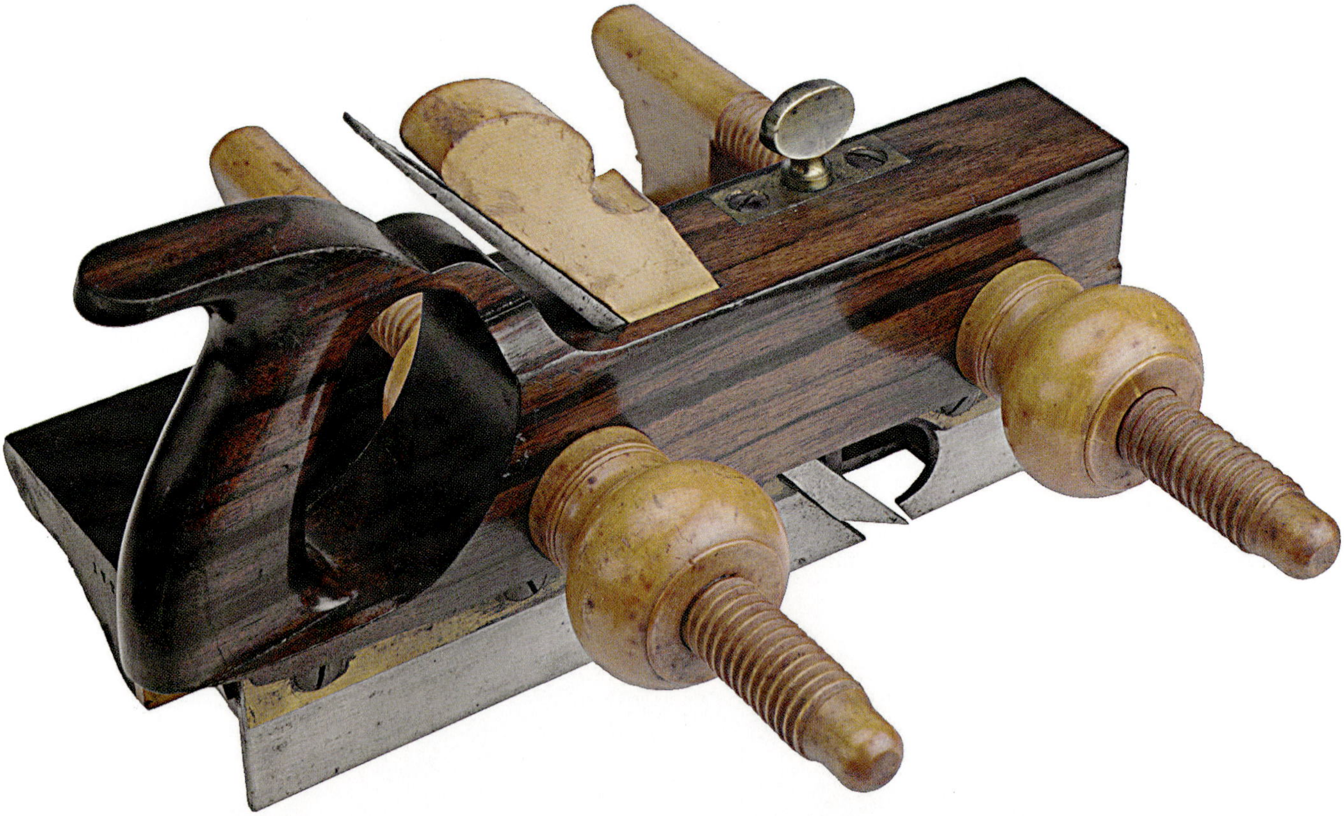

J. H. Lamb

Handled Macassar ebony screw-arm plane with boxwood arms, nuts and wedge, boxed fence and nickeled depth screw lock. *

(From the collection of Don Rosebrook)

James Lamb was a New Bedford, Mass. planemaker who was active from 1869 to 1874, but it is unclear how much of that time he operated under the name of J. H. Lamb. Pollak believes that he was a part of J. & W. Lamb for part of this time and that he may have been a partner in Lamb and Brownell.

Note the distinctive shape of the boxing. Many varieties of boxing joints are exhibited on the planes in this book. It seems strange that the depth screw lock is nickeled. Perhaps it was an after-market addition, but this plane has never been used and retains all of its original lacquer. Perhaps it was the only type in stock when the plane was made.

J. H. Lamb

Unsigned handled mahogany screw-arm plow with boxwood arms and nuts, boxed fence and a depth screw lock.

(From the collection of Don Rosebrook)

Mahogany was used for plow planes much less frequently than rosewood, certainly not because of the lack of beauty, as this plane demonstrates. The shape of the skate nose as well as the nuts and the various moldings confirm that this is indeed a J. H. Lamb plane.

Nearly all of the mahogany planes that the authors are familiar with utilized Honduras mahogany. This is strange because all or nearly all of the levels made in this part of the US during the mid to late 1800s utilized Cuban mahogany. Cuban mahogany was the premier mahogany for 250 years until wastage destroyed its commercial availability.

J. H. Lamb

Handled rosewood screw-arm plow with boxwood wedge, arms and nuts, boxed fence, and depth stop lock. *

(From the collection of Don Rosebrook)

Note again the distinctive skate nose and the molding features of the plane. These planes also use screws that are slightly domed. The standard for plane makers was flat headed slotted screws. A very few used domed screws or round headed screws. One plane maker, DeForest, used flat screws on most planes but used round headed screws whenever the screw was to be nickel plated. The presence of round headed screws on a DeForest plane is an indication that the metal parts were once nickeled and that it was one of the better quality planes from this maker.

J. H. Lamb

Handled mahogany screw-arm plow with boxwood arms and flat nuts and ivory and boxwood sandwich nuts, depth stop lock, and silver mounting. Mark does not show location, and is not recorded in Pollak as such, but still should rate at least *.

(From the collection of Patrick Leach)

Having stated that nearly all of the planes made of mahogany utilized Honduras mahogany, this plane is presented as being made of Cuban mahogany and is a fine example of the beauty of that wood.

Lamb & Brownell

Handled rosewood screw-arm plow with brass arms, boxwood nuts, boxed fence, depth stop lock, and silver mounting. *

(From the collection of Don Rosebrook)

Lamb & Brownell was reported by Pollak to have been operating in New Bedford, Massachusetts in 1871. No other details of the operation were given except to speculate that the planemakers may have been Charles Lamb, James H. Lamb, and/or William Lamb.

The use of brass arms is rare, although another example is given in this book (see Sandusky Tool Co.). In this case, the shoulders of the arms are made with a nicely contoured sweep instead of the characteristic molding form. It is not possible to determine if these arms are original to the plane.

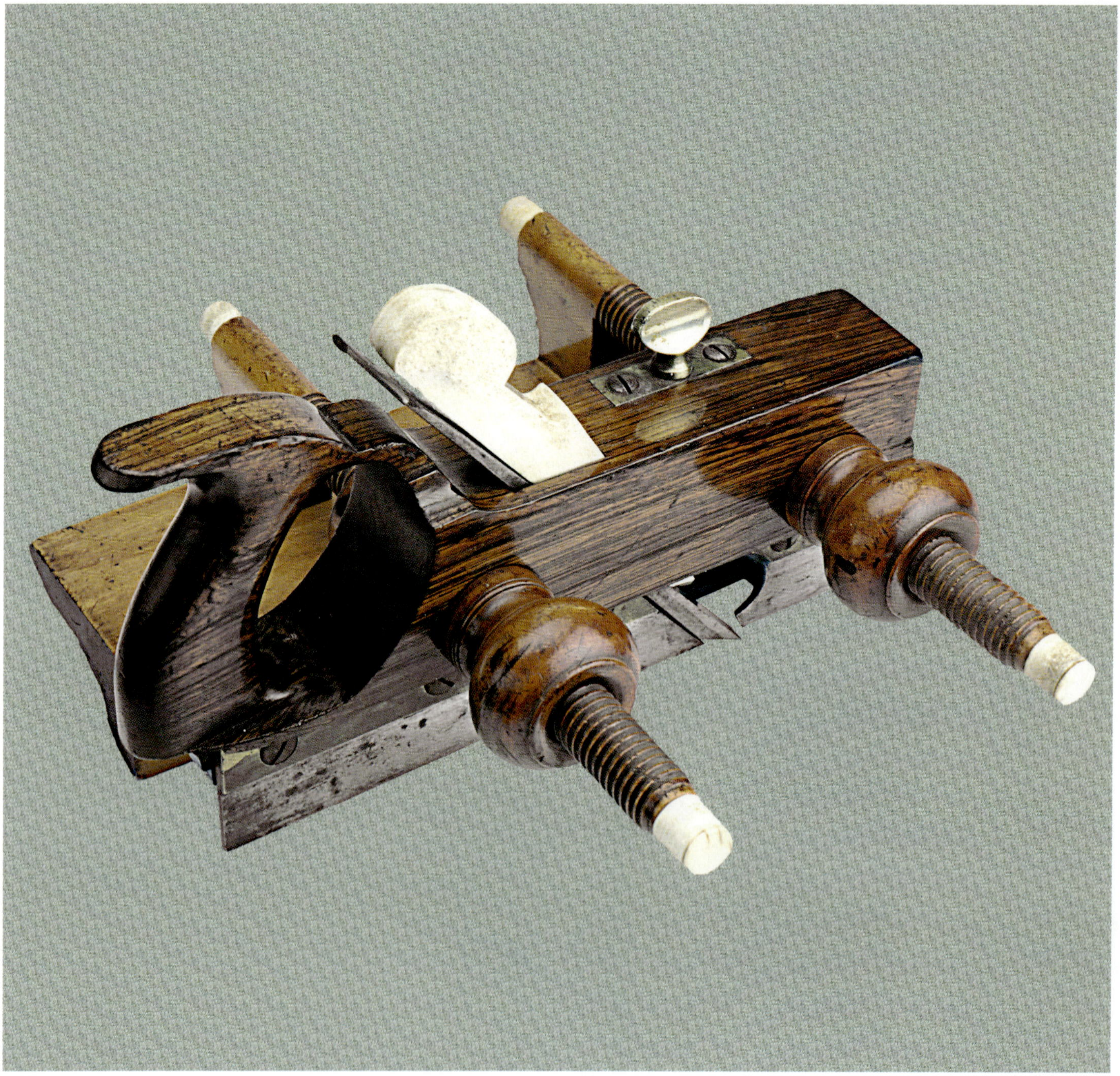

Lamb & Brownell

Handled rosewood screw-arm plow with boxwood fence, arms and nuts, ivory wedge and tips. Brass inserts in the fence receive the screws. Screws are slightly domed. *

(From the collection of Don Rosebrook)

The moldings and the nuts are virtually indistinguishable from those on planes marked J. H. Lamb.

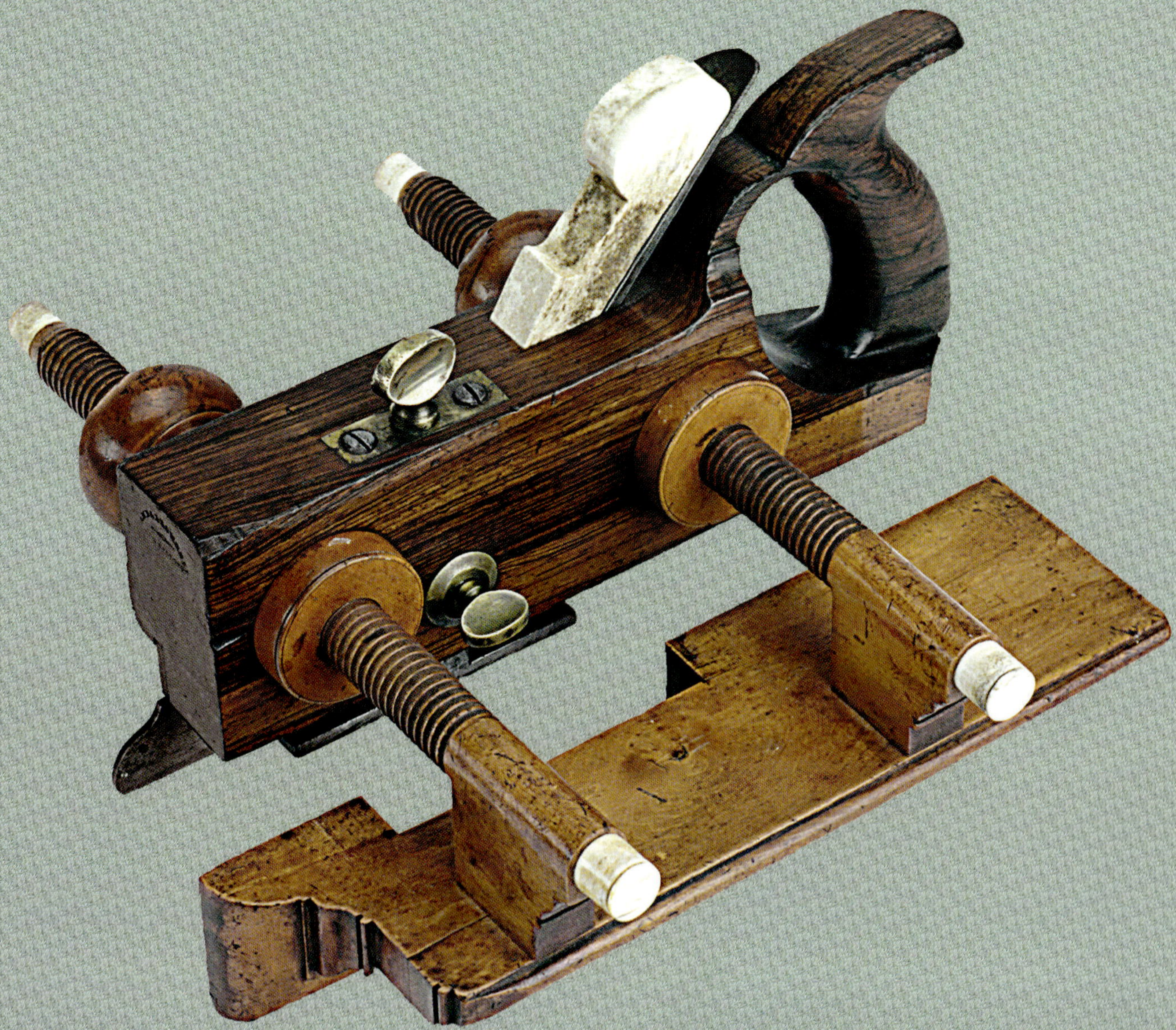

Although, most of the ivory used on planes was elephant ivory, the ivory on this plane is thought to be walrus ivory due to the many dark mottled areas. Walrus ivory has such areas at the center of the tusk.

Lamb & Brownell

Handled rosewood screw-arm plow with boxwood fence, arms & nuts, depth stop lock, silver mounting, and full readable label. *

(From the collection of Patrick Leach)

The presence of the full label in nearly unscratched condition is remarkable. The plane, of course, has seen little if any use. Once again, it is possible to see that the molding features and nuts of the Lamb & Brownell planes are indistinguishable from those of the J. H. Lamb planes. This means that the unsigned mahogany plane earlier identified as a Lamb could be a Lamb & Brownell. However, the shape of the wedge and the skate nose tend to favor J. H. Lamb as the maker.

Lamb & Brownell

Handled boxwood screw-arm plow with ivory wedge and depth stop lock screw. *

(From the collection of Don Rosebrook)

The ivory wedge seems to be slightly too narrow for the throat of the plane. It may be an aftermarket addition.

Multiform Molding Plane Co.

Beech screw-arm plow with boxwood nuts and arms of an unfamiliar wood (perhaps W. Indian Satinwood). The detachable handle is beech. The plane is marked "MULTIFORM MOLDING PLANE CO. Patented Aug. 29, 1854".

(From the collection of Don Rosebrook)

Note the difference in the handles on this plane and on the ebony plane, shown on the next two pages. The handle used on this plane leaves room for a full sized blade. The handle on the ebony plane requires a short blade because a long blade will not clear the handle.

Multiform Molding Plane Co.

Ebony screw-arm plow with boxwood arms and nuts, and silver trim. The detachable handle is ebony. The plane is marked "MULTIFORM MOLDING PLANE CO. patented Aug. 29, 1854".

(From the collection of Don Rosebrook)
Photographs this page and next by Larry Weaver.

The Multiform Molding Plane Co. was established in Boston in 1856 and operated until 1858. Thomas Worrall used the company to manufacture planes according to his patents. The August 29, 1854 patent was for planes with interchangeable soles.

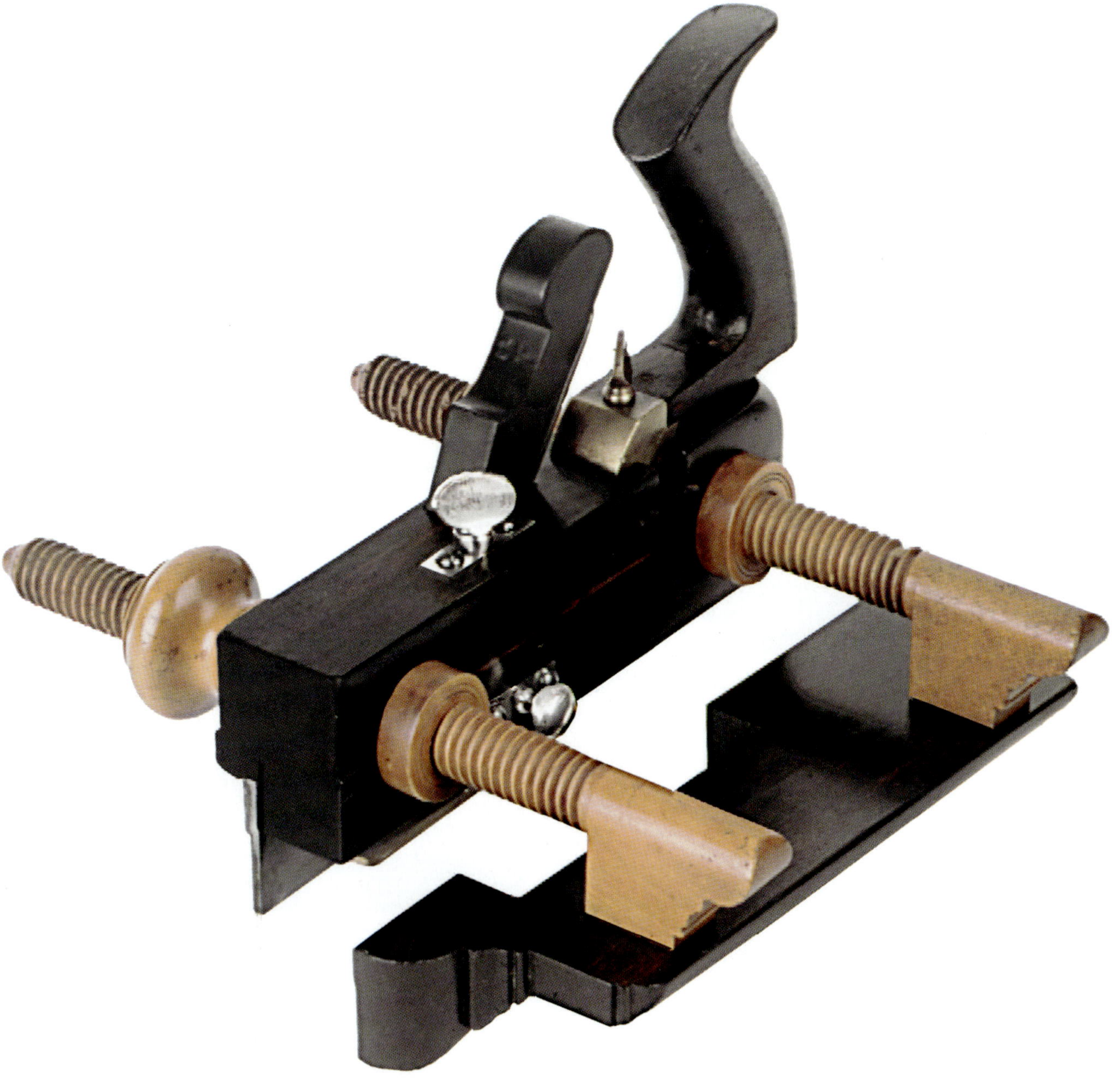

No other ebony plow planes are known by this maker. Add to this the silver (not nickel) trim and it appears as though this plane is a special order plane of some type, perhaps a presentation piece.

F. Nicholson

Unhandled yellow birch screw-lock slide arm plane with depth stop lock. Marked "F. NICHOLSON LIVING IN WRENTHAM". **

(From the collection of David Englund)

The owner's mark on this plane is for B. Mann. B. Mann most likely stands for one of the grandchildren of Samuel Mann, a minister in early Wrentham, Massachusetts. They would have been contemporaries of Nicholson and were named Benjamin, Beria and Beriah.

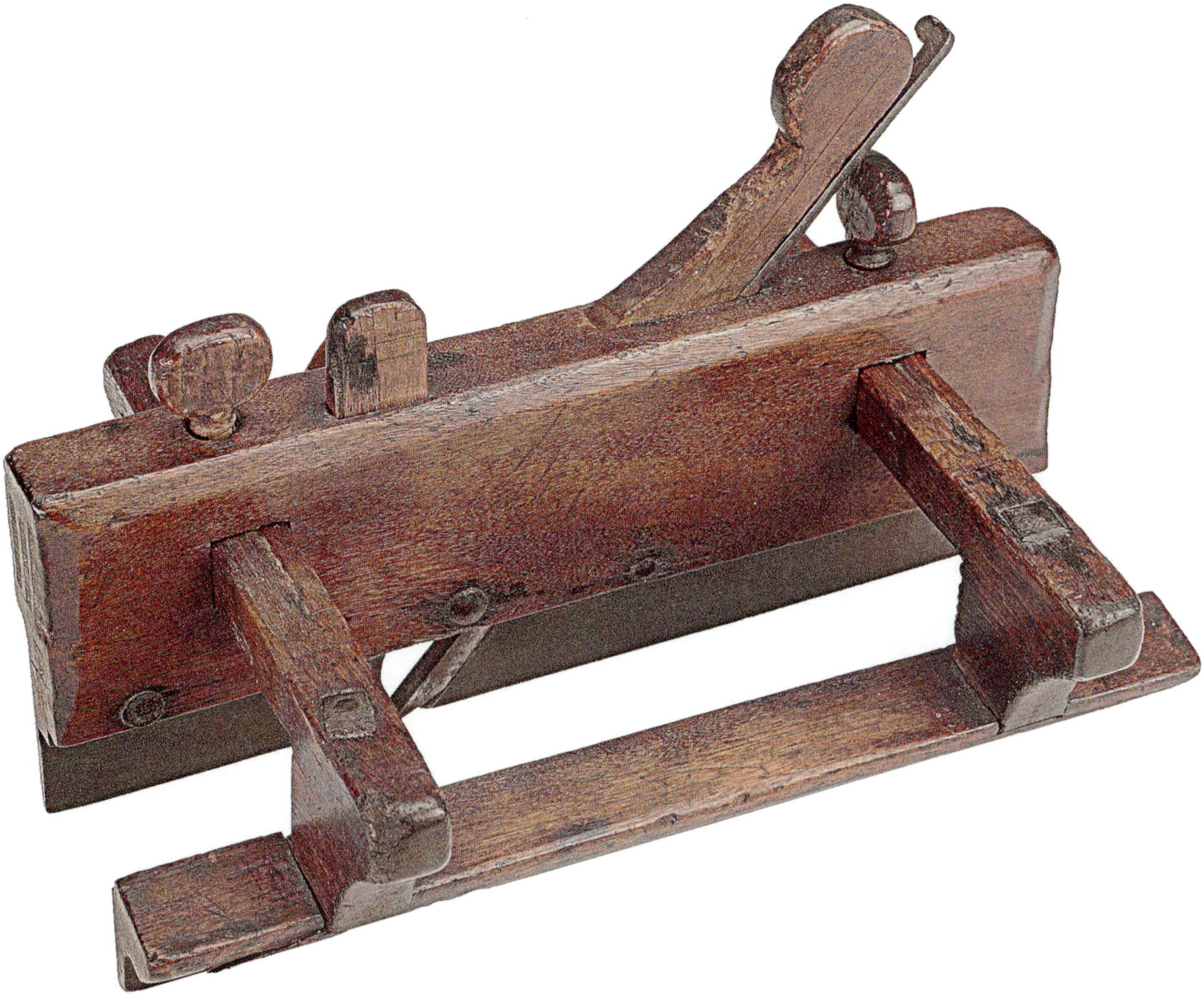

Francis Nicholson is believed to be America's first documented plane maker. According to Pollak, Francis was listed as a joiner in Rehoboth, Massachusetts. in 1716. He moved to Wrentham, Massachusetts in 1728. It is possible that he made planes before he moved to Wrentham but there is no evidence to support or disprove that premise. He lived from 1683-4 to 1753. Another early American planemaker, Cesar Chelor, was Nicholson's black slave. Chelor was freed upon Nicholson's death.

F. Nicholson

Unhandled yellow birch wedge-lock slide-arm plow. Marked "F. Nicholson." New mark now called the D mark. *****

(From the collection of Michael Humphrey)

This plane is believed by some to be the earliest Nicholson found to date. The mark is crude and the workmanship is also somewhat crude. The plane has seen a lot of use and shows the wear. Chamfers on this plane are extremely broad.

S. A. Piper

Handled beech patented plow with mahogany wedge. Plane patented June 28, 1881, in Upper Newton Falls, Massachusetts. ****

(From the collection of Steve Dice)

S. A. Piper apparently made planes for, or with, T. W. Piper because planes with both their marks are known. They may have been related; however, based on the information in Pollak, it would appear that S. A Piper worked much later than either T. W. or A. M. Piper, both of whom also worked in Newton, Massachusetts.

This plow has a fence that swings on a pair of pivoted brass bars attached both to the fence and the body. A steel bar is free to move through the body, and when the lock screw on the top of the body is turned, the fence is locked in place. Both the fence and the body have round mortises to receive the fitting from the other side of the pivoted bar.

E. Smith

Unhandled beech screw-lock slide-arm plow, unplated.

(From the collection of Don Rosebrook)

Ezekiel Smith worked as a plane maker from 1823 to 1874. He began in Rehoboth, Massachusetts. but moved to Smithfield, Rhode Island in 1850. In 1857 he moved to Worcester, Massachusetts and stayed there. He was known to have a fairly large output.

This plane is an inexpensive model because the skate is not plated. It also has characteristics that imply it was a later production, possibly from the R.I. shop or from Worcester.

I + Iones

Unhandled birch screw-lock slide arm plow. ****

(From the collection of Mickey Holmes)

Jethro Jones was a black planemaker who worked from 1764 to 1767.

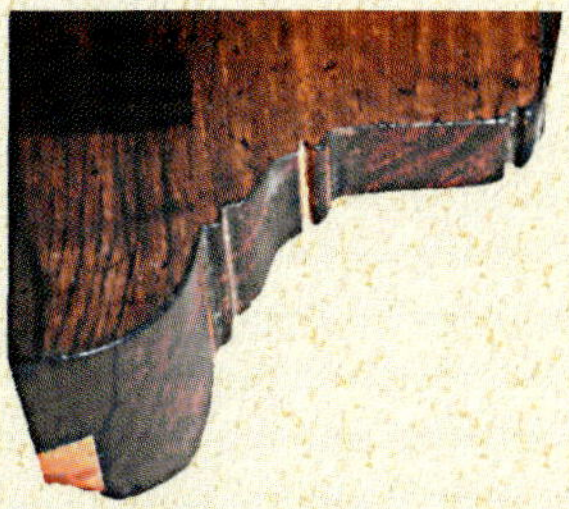

Taber Plane Co.

Handled rosewood screw-arm plow with boxwood arms and nuts, depth stop lock, and silver mounting.

(From the collection of Don Rosebrook)

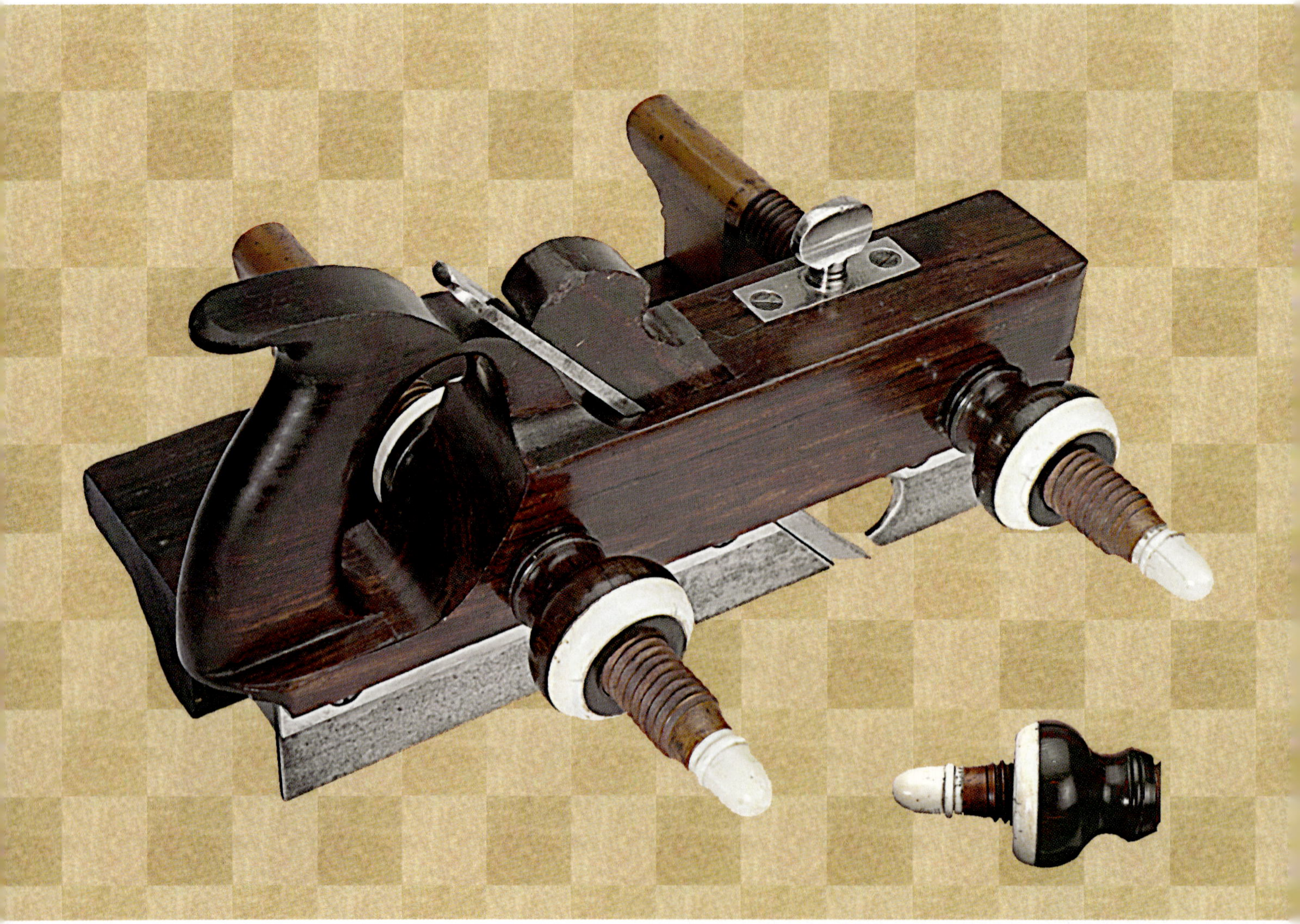

Taber Plane Co.

Handled rosewood screw-arm plow with boxwood arms and ivory/rosewood sandwich nuts, depth stop lock, and silver mounting.

(From the collection of Don Rosebrooki)

These rosewood/ivory sandwich nuts and those on the Lamb plow planes are undoubtedly patterned after the Greenfield Tool Co. nuts. However, the two-piece sandwich construction here is different.

The Taber Plane Co. was founded in 1866, succeeding J. & W. Lamb. Its officers were James and William Lamb. The company ceased operations in 1872. Thus, it is no accident that the forms and styles of nuts and moldings were similar among the various Lamb planemaking operations and the Taber Plane Co.

Note that the flat nuts are also of sandwich construction. To the authors' knowledge, no Greenfield plow utilized flat nuts constructed like this.

Taber Plane Co.

Unhandled boxwood screw-arm plow with large nuts of brass and depth stop lock.

(From the collection of Don Rosebrook)

The "large" nuts are flat nuts in this case and have a smaller diameter inner part. The outer flange is knurled. The mark also contains a sunburst that appears to be original.

J. Taber

Unhandled beech screw-lock slide arm plow with depth stop lock. ****

(From the collection of Frank Kosmerl)

The operating period and location of J. Taber are unknown. From the characteristics of the plane, it can be estimated that the operating period was around 1790. It is also speculated that he was associated in some way with Nicholas Taber of New Bedford, Massachusetts.

There is a metal plate fastened to the face of the fence as a guard to prevent wear.

W. H. Taber

Unhandled beech screw-arm plow with birch fence and arms and boxwood nuts and wedge. Unlisted mark, which consists of four W. H. Taber marks applied in a square, with both vertical marks facing left and the bottom mark being upside down. Probably *** or ****

(From the collection of Peter Hathaway)

The nuts on this plow are of a design quite different from that of other manufactured plow planes.

Wing H. Taber began making planes in Fairhaven, Massachusetts in 1833. He is said to have been making planes in New York City from 1836 to 1843, and thereafter in Lowell, Massachusetts until 1866. While he is not related to the Tabers of Taber Plane Co., the Taber Plane Co. acquired his patent and produced bench planes according to its specifications.

It is unusual to find a plow plane with the arms mortised into the fence from a planemaker working this late, not to mention finding such arms on a screw-arm plow.

I: B: Waltons

Unhandled beech wedge-lock slide-arm plow with unique depth stop. ****

(From the collection of Robert Wheeler)

I: B: Waltons was really John and Benjamin Walton.They made planes in Reading, Mass. until 1771. Pollak does not give a beginning date. The threaded bar functioning as a depth stop is unique to this maker and the fence with a molding at both ends is unusual. Such a fence was seldom seen again until adopted by the Auburn Prison plane makers one hundred years later.

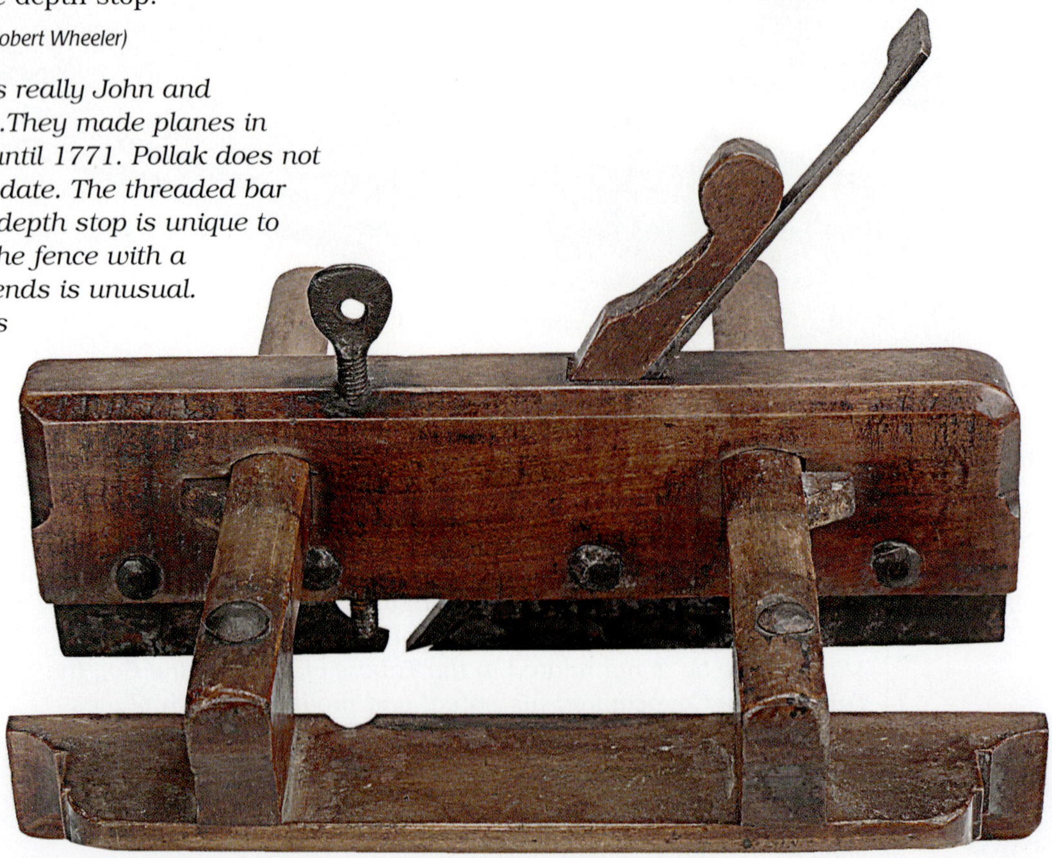

B. Webb

Unhandled beech screw-arm plane with cast iron arms and brass nuts, depth stop lock.

(From the collection of Bill Curtis)

Pollak does not contain an entry for B. Webb. The wooden parts of this plane, however, appear very well done and may have been manufactured.

Peeler

Unhandled beech screw-lock, slide-arm plow with birch fence and oak wedge and thumb screws. ***

(From the collection of Mickey Holmes)

This mark is believed to belong to Alender Peeler of Vernon, Vt. His working period is estimated to be 1800-1810.

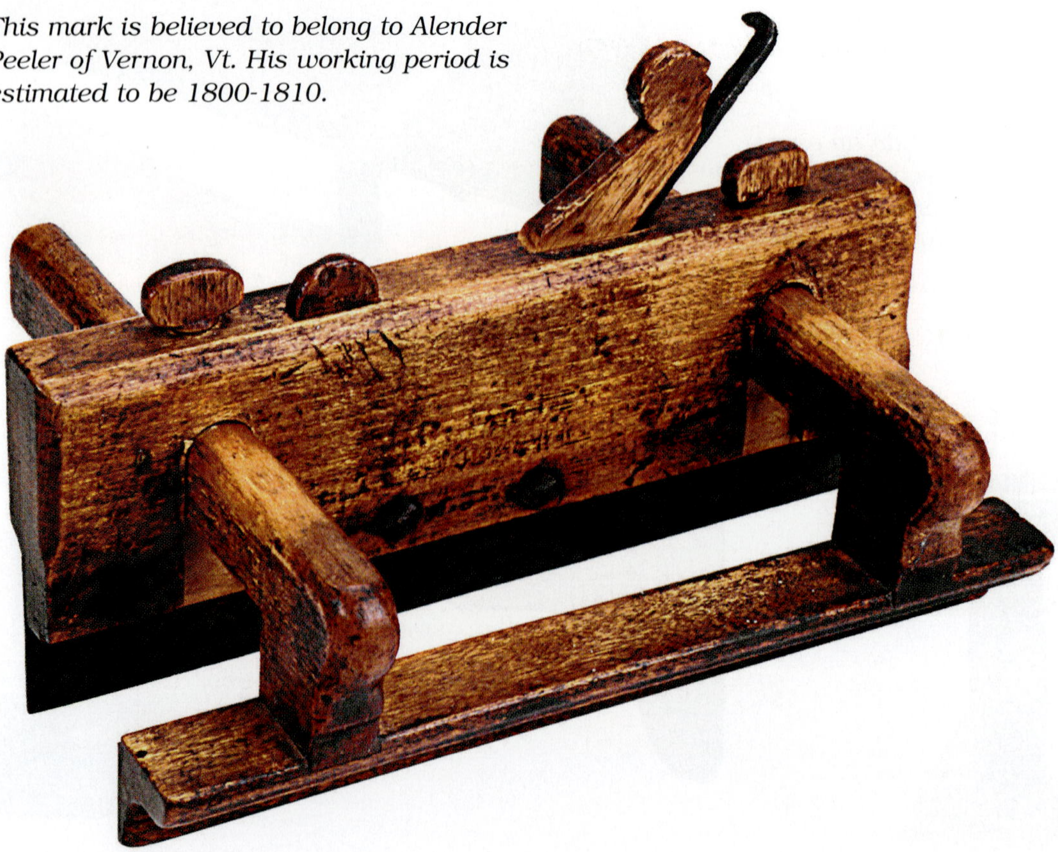

D. P. Sanborn

Unhandled beech screw-lock, slide-arm plow, unplated, with depth stop lock. B1 mark **

(From the collection of Kenneth Thomas <<< Don Rosebrook)

This plane is in mint condition. It has never been used and not even handled enough for body oils to darken spots on the wood. Since Sanborn worked in an era where all but the low end planes had plated skates, this must have been an inexpensive plane.

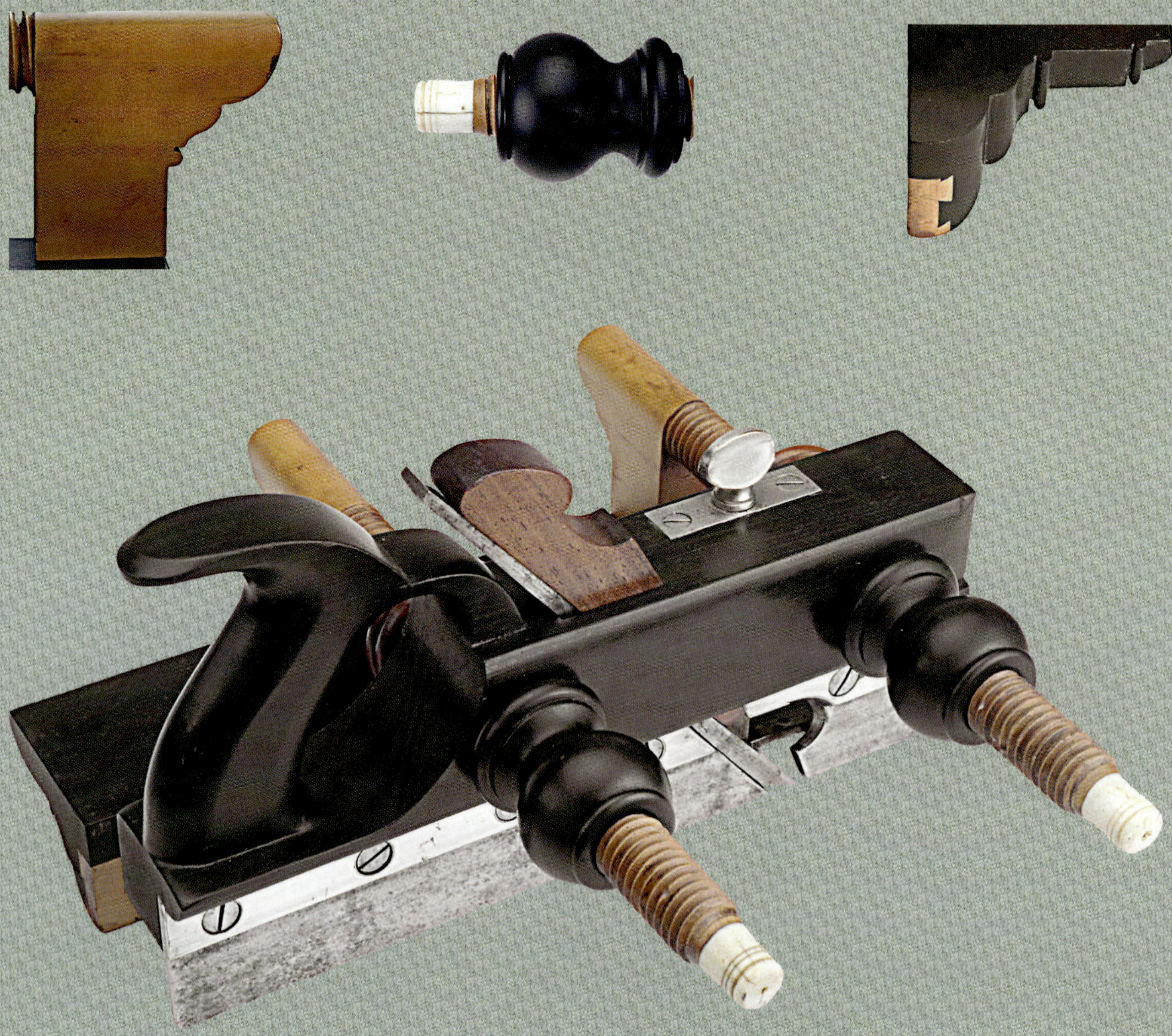

D. P. Sanborn

Handled ebony screw-arm plow with ebony large nuts, boxwood arms and boxed fence, rosewood wedge and flat nuts, two ivory tips, and silver mounting. (E mark ***)

(From the collection of Don Rosebrook)

This is a remarkably attractive plane and has been very little used. Note the boxing on the fence — the tail of the dovetail joint has been formed on the fence, which is unusual.

David Page Sanborn was listed as a planemaker in Littleton, N. H. before 1840. He went to Worcester, making planes as a partner in Sanborn & Gouch and perhaps on his own. He returned to Littleton about 1850, making planes there until 1871.

C. Warren

Unhandled beech screw-arm plow with boxwood arms and nuts. This is a left-handed plane.

(From the collection of James Leamy)

Left-handed planes must be made that way. This is obvious because a standard fence is pointed only on the forward end and, thus, is not reversible. (There is an exception to this in that many of the New York prison-made planes are pointed on both ends.) The other characteristic is the position of the mortise for the depth stop. The fence on a left-handed plane is the mirror image of the fence on a standard plane.

C. Warren

Unhandled rosewood screw-arm plow with boxwood fence, arms, nuts, wedge and depth stop lock.

(From the collection of Rick Benze)

Cyrus Warren made planes in Hudson, New Hampshire from 1837-1857 and in Nashua, New Hampshire from 1858-1875.

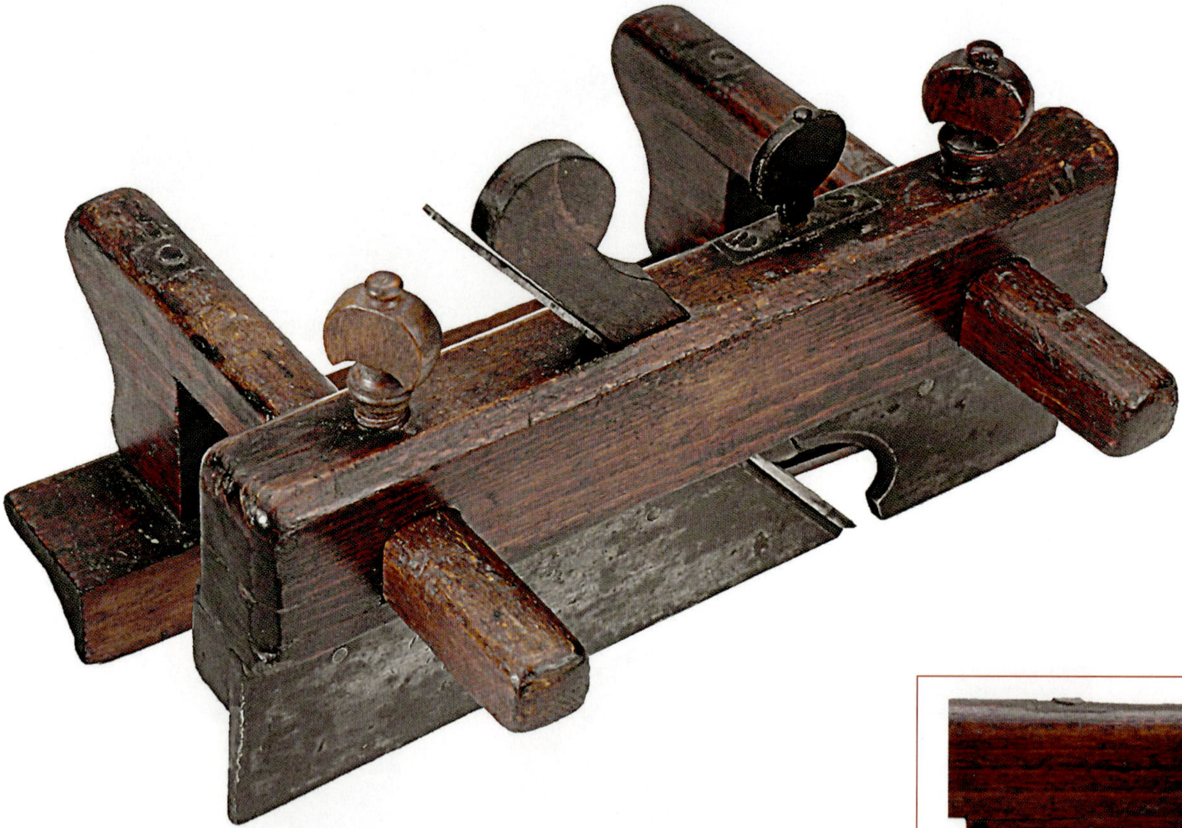

E. Clifford

Unhandled beech screw-lock, slide-arm plow. ****

(From the collection of Robert Wheeler)

Ebenezer Clifford worked in Kensington, New Hampshire from 1772-1793 and in Exeter, New Hampshire from 1793-1808.

The skills of the maker are obvious in the look of the thumb screws, both metal and wood. The depth stop is unusual in that the base has two support posts, where one is the norm. Also, the maker affixed the skate using screws from the opposite side of the body, meaning that the holes in the skate had to be threaded.

I. S. Battey

Unhandled beech screw-lock, slide-arm plow with boxwood depth stop and depth stop lock.

(From the collection of Robert Bills)

Battey was a Providence, Rhode Island planemaker who worked during 1841-1855.

P. H. Manchester

Unhandled rosewood screw-lock, slide-arm plow with depth stop lock, brass collars on the ends of the arms and boxwood wedges in the ends of the arms. *

(From the collection of Robert Bills)

Pardon Manchester was a Providence, Rhode Island planemaker who worked from 1841 to 1858.

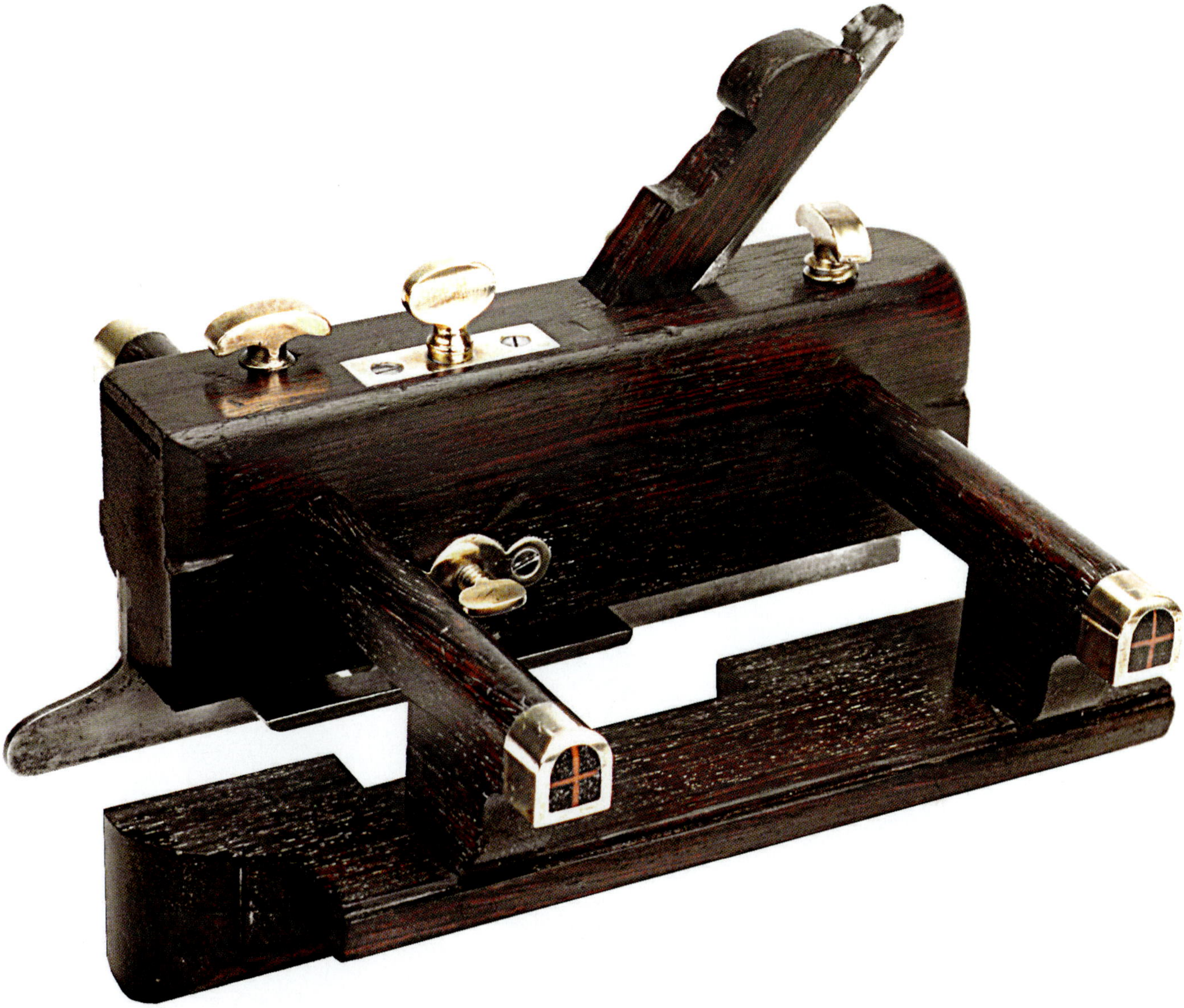

Two things stand out about Rhode Island planes. First, many of the planes employ brass collars on the ends of the arms and contrasting color wedges in the ends of the arms to secure the collars. Second, many of the planes have skate noses that are not square, while square nosed skates are the norm in the rest of New England.

P. H. Manchester

Handled beech screw-lock, slide-arm plow with rabbets on ends of arms. *

(From the collection of Robert Bills)

It appears that this plane was, at one time, fitted with pieces to extend the arms. The arms, as shown, are quite short (9½") for such a long plane (12½").

P. H. Manchester

Unhandled boxwood screw-lock, slide-arm plow with depth stop lock, brass collars on the ends of the arms and ebony wedges in the ends of the arms. *

(From the collection of Robert Bills)

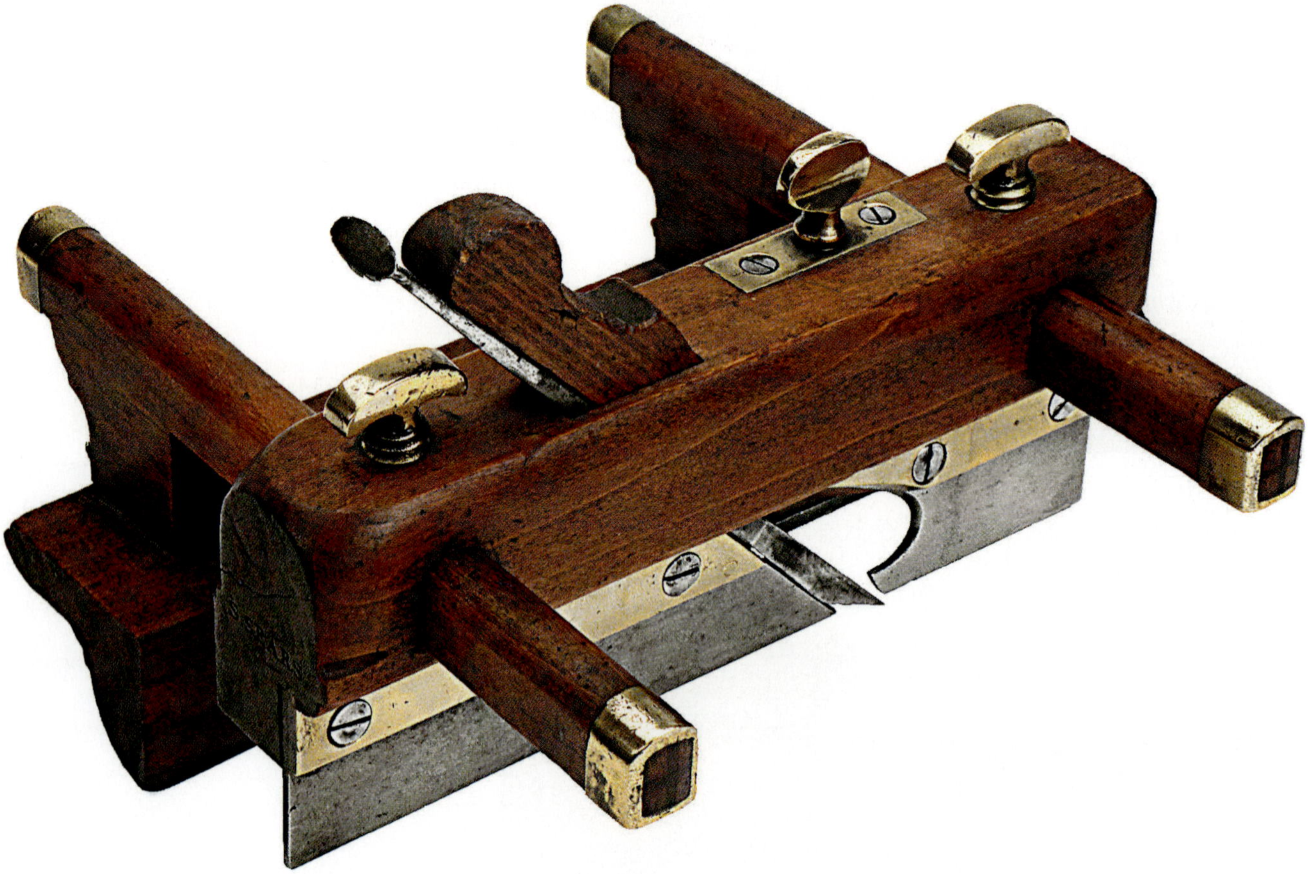

J. W. Pearce [J. W. Pirce]

Unhandled beech screw-lock, slide-arm plow with brass collars on the arm tips, rosewood wedges in the arm tips, and depth stop lock (missing thumb screw). D1 mark **

(From the collection of Robert Bills)

According to Pollak, Jonathon Pearce worked as a planemaker in Fall River, Massachusetts from 1840 to 1852 and in Providence, Rhode Island from 1853-1879. Many of his planes are imprinted with the date that they were produced.

J. W. Pearce

Unhandled mahogany screw-lock, slide-arm, plow with rosewood fence, mahogany arms, boxwood wedge, and depth stop lock. Skate nose turned up and additional signature on the skate plate. New mark on nose — like B mark w/o Prov. R. I.

(From the collection of Robert Bills)

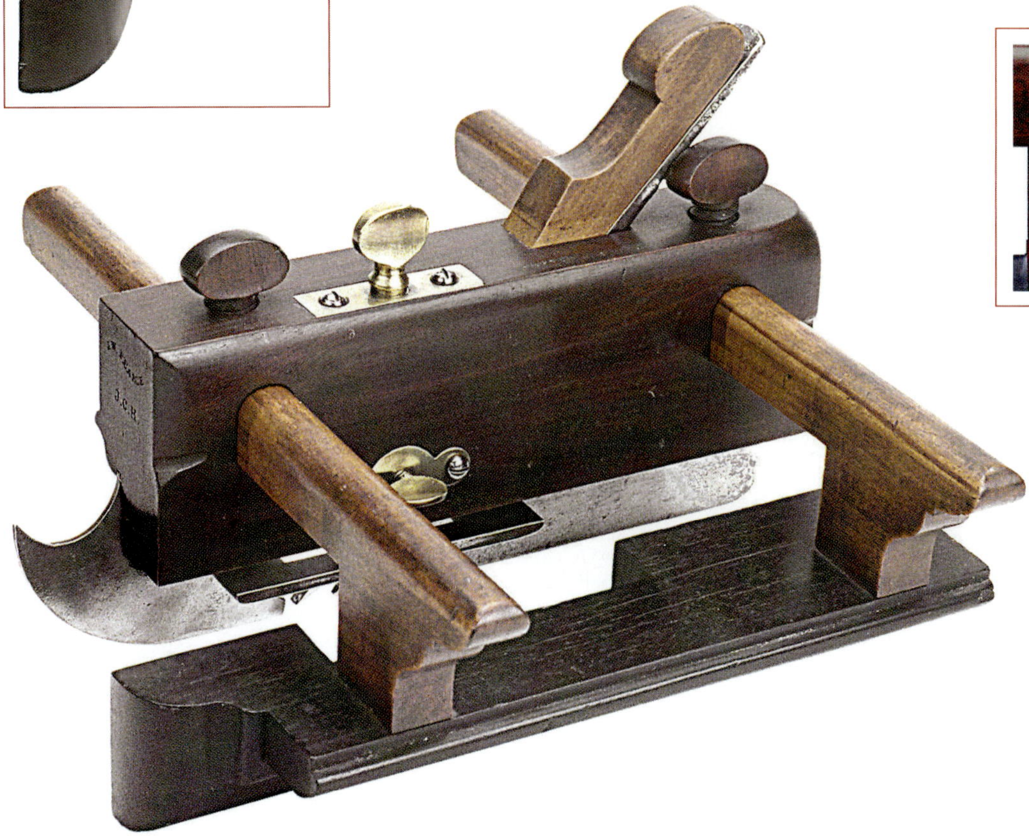

J. W. Pearce

Unhandled mahogany screw-lock, slide-arm plow with ebony fence, fruitwood arms and wedge, and depth stop lock. New mark on nose — like B mark w/o Prov. R. I.

(From the collection of Don Rosebrook <<< Robert Bills)

J. Edwin Child

Unhandled beech screw-lock, slide-arm plow with turned up skate nose. B mark **

(From the collection of Peter Hathaway)

J. Edwin Child was a Providence, R. I. planemaker who worked during 1850-1880. He is known to have done contract work for the Greenfield Tool Co.

I. E. Smith

Unhandled beech screw-lock, slide-arm plow with round arms. A mark **

(From the collection of Robert Bills)

Ira Smith was a planemaker living in Smithfield, Rhode Island and working in Providence, Rhode Island. He died in an explosion in Providence, in 1857, at age 27. He was the son of E. Smith (which see).

Hermon Chapin

Handled applewood patent adjusting plane — first model; with boxwood center screw and lock, boxed fence and depth stop lock. E. M. Chapin & Solon Rust patent March 31, 1868. Model #239½.

(From the collection of Don Rosebrook)

H. Chapin became the largest of the Connecticut planemakers. Union Factory, on the North Branch of the Farmington River, was founded in 1826 and came under the sole control of Hermon Chapin in 1828. The Chapin family operated this business under various names until 1901, when it was merged with L. C. Stephens & Co to form the Chapin-Stephens Co. Regardless of the company name during this 75-year period, the planes were marked as "H. Chapin."

The patent of E. M. Chapin (Hermon Chapin's son) and Solon Rust was granted on March 31, 1868 and was first offered in the 1869 catalog of Sargent & Co. These planes were offered in boxwood, beech, applewood, and rosewood and the latter three were offered with and without boxed fences. According to the patent, the object was to "... construct a joiner's plow ... provided with adjustable fences in such a manner that the fence-guides and screws will not extend through the body ...." There was no discussion of the idea of self-regulating or keeping the fence parallel to the skate by use of a single adjustment, as is commonly thought to have been the rationale. The peg shown beside the wedge serves to retain the threaded arm in the body when the arm is turned.

Hermon Chapin

Handled beech patent adjusting plane — second model; unplated, with depth stop lock. E. M. Chapin & Solon Rust patent March 31, 1868. Model #239½.

(From the collection of Don Rosebrook)

This is generally accepted as the second model of the Chapin Rust plow plane. It uses the same approach as the first model, except that the arms are now round and the center screw is now cast iron. This particular plow has obviously been used very little and the hardware store notations are quite visible on the top of the body near the front. This model of the plow was not shown in any Chapin catalog.

Hermon Chapin

Handled beech patent adjusting plane — third model; rosewood arms with a shallow V groove on their bottom side, iron bridle, unplated, with depth stop lock. E. M. Chapin & Solon Rust patent March 31, 1868. Model #236$^{1}/_{4}$.

(From the collection of Don Rosebrook)

This plane is generally thought to be the third model of the Chapin-Rust patented plow. The rosewood arms have a shallow V-groove cut into their bottom edge. The cast iron supports beneath the arms form a ridge under each arm. This feature is intended to keep the fence moving in a straight line during its adjustment. The patent adjusting planes are really all 3-arm bridle plows. The plane on the next page goes to the next logical step, which is the elimination of the 3rd arm to form a true bridle plow.

Hermon Chapin

Handled bridle plane — rosewood body, boxwood fence and wedge, ebony arms with brass tips and rounded nose on the skate. No model number. Depth stop lock and cast iron bridle.

(From the collection of Don Rosebrook)

There appears to be only one known example of this particular plane, regardless of the type of woods involved in its manufacture. This means that it is probably a prototype. A pattern matching the skate was found in Solon Rust's tool chest. Note that the V-groove in this plane is in the bottom part of the bridle mechanism. In the third model of the Chapin-Rust plane, the V-groove is on the bottom of the arm. This plane resembles the Mathieson bridle plows.

Hermon Chapin

Handled boxwood screw-arm plane with cast iron arms, large brass nuts, rosewood flat nuts, and beech wedge.

(From the collection of Mickey Holmes)

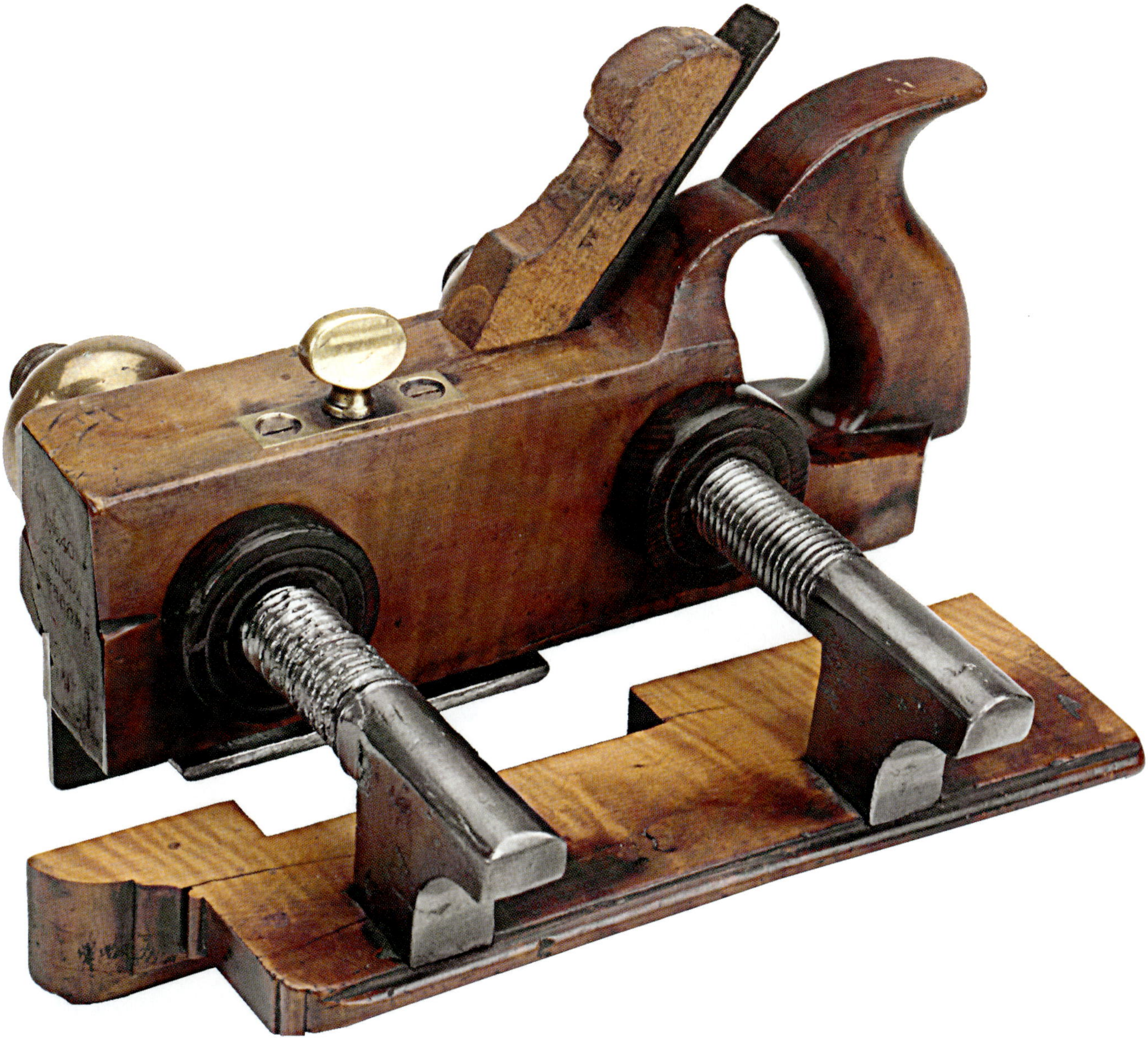

The use of brass nuts has been observed on planes made by other makers; however it is a rare feature. The use of cast iron arms is still more rare and this is the only example known to the author. The nuts appear to have the proper shape; however, the arms do not have the Chapin detail at their shoulders.

Hermon Chapin

Handled boxwood screw-arm plane with rosewood nuts. Model #240. Over-stamped Way & Sherman.

(From the collection of Don Rosebrook)

It is natural to think that if a boxwood plane has rosewood nuts, then the nuts must have been replaced by an earlier owner. However, in this case, the plane is identical to the plane shown in Plate O of Ken Roberts' book. Both planes carry a Way & Sherman overstrike. Both planes have some sapwood in the nuts and both planes show some striped figure in the boxwood. This implies that Way & Sherman, a New York City hardware store, requested the distinctive feature. The features are stronger in the Roberts' plane.

Hermon Chapin

Handled boxwood screw-arm plane with depth stop lock. Model #240.

(From the collection of Don Rosebrook)

Two features of H. Chapin planes that changed over the years were the shape of the nuts and the tips of the arms. Some arm tips were buttons like these, while others were ball shaped; and still others were blunt.

Hermon Chapin

Handled rosewood screw-arm plane with boxwood arms and nuts and wedge, boxed fence, and depth stop lock. Model #240½.

(From the collection of Don Rosebrook)

The similarities between Chapin nuts of this style and the nuts found on DeForest planes may be due to the fact that Solon Rust worked at DeForest prior to joining Chapin at Union Factory.

Hermon Chapin

Handled mahogany screw-arm plane with boxwood arms and nuts. Model #240. Overstamped H. Perkins & Co., Winstead, Ct.

(From the collection of Peter Hathaway)

Mahogany planes are not commonly found, although most plane makers of the mid-1800s did make a mahogany plane.

L. DeForest

Handled ebony screw-arm plow with boxwood arms and nuts and silver mounting.

(From the collection of Don Rosebrook >>> ??)

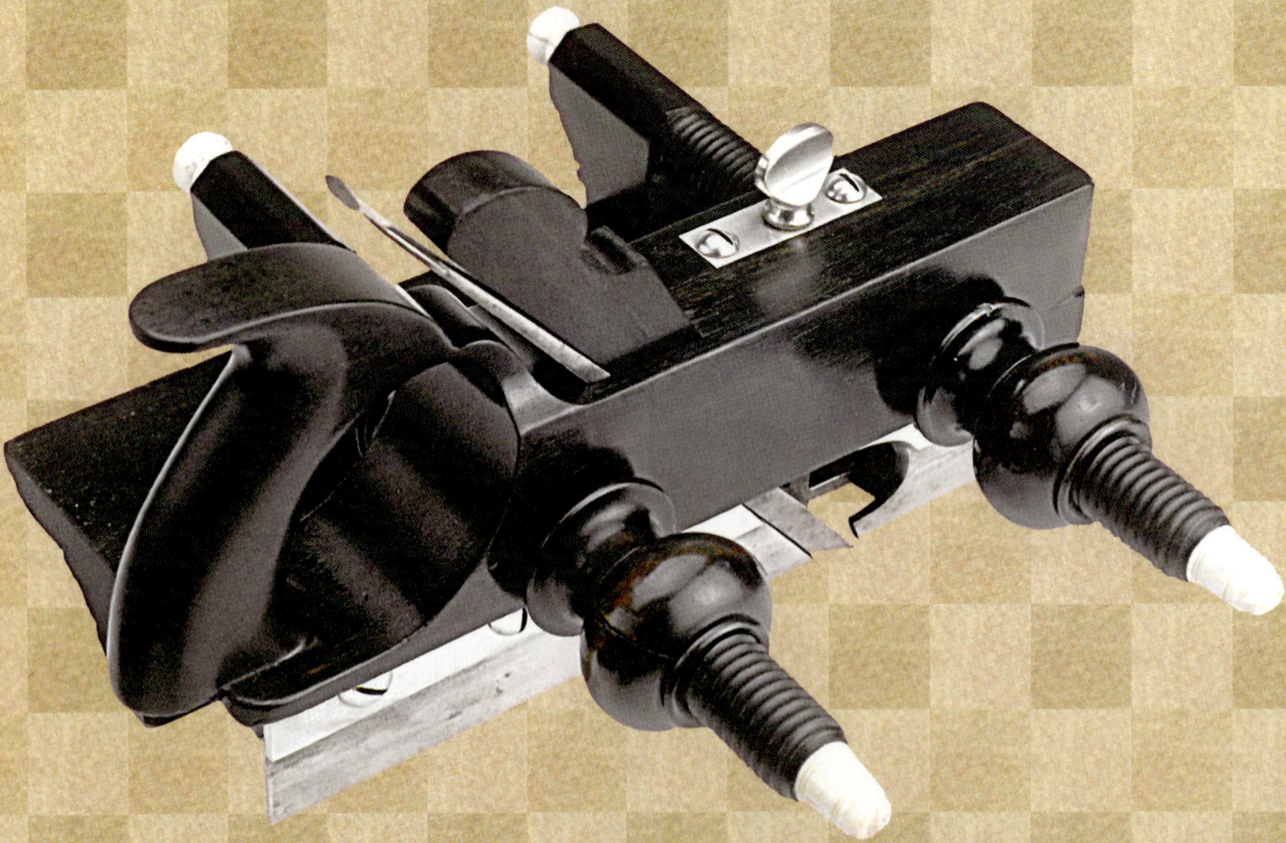

L. DeForest

Handled ebony screw-arm plow with ebonized boxwood nuts, ivory tips and silver mounting.

(From the collection of Don Rosebrook)

DeForest made planes in Birmingham, Connecticut (now Derby) from 1850 - 1857 and 1859 - 1860. He was part of L. DeForest & Co.

"Silver mounted" or "silver mounting," means nickel plated or plated with nickel-silver (German silver), in most cases. Some examples of actual silver plating are known.

This plane, as is the case with most DeForest plows, is unmarked. It is near the top of the line because it utilizes ebony for the arms, ebonized boxwood for the nuts and is silver mounted. A large number of DeForest planes are silver mounted and all of these planes use round headed screws such as those visible above. Flat headed screws were the norm for the industry.

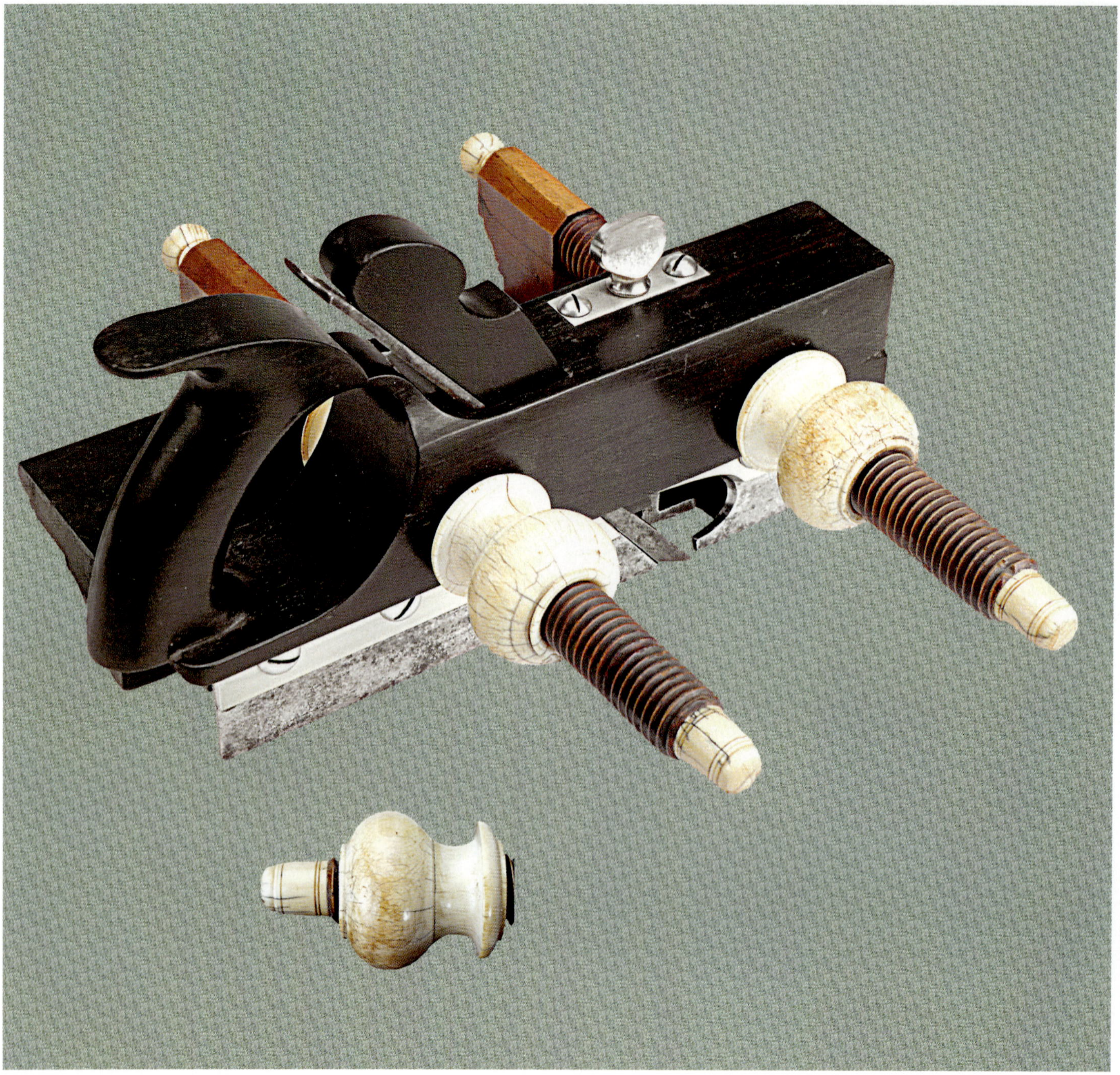

L. DeForest

Handled ebony screw-arm plow with boxwood arms, ivory nuts and ivory tips, and silver mounting.
(From the collection of Don Rosebrook)

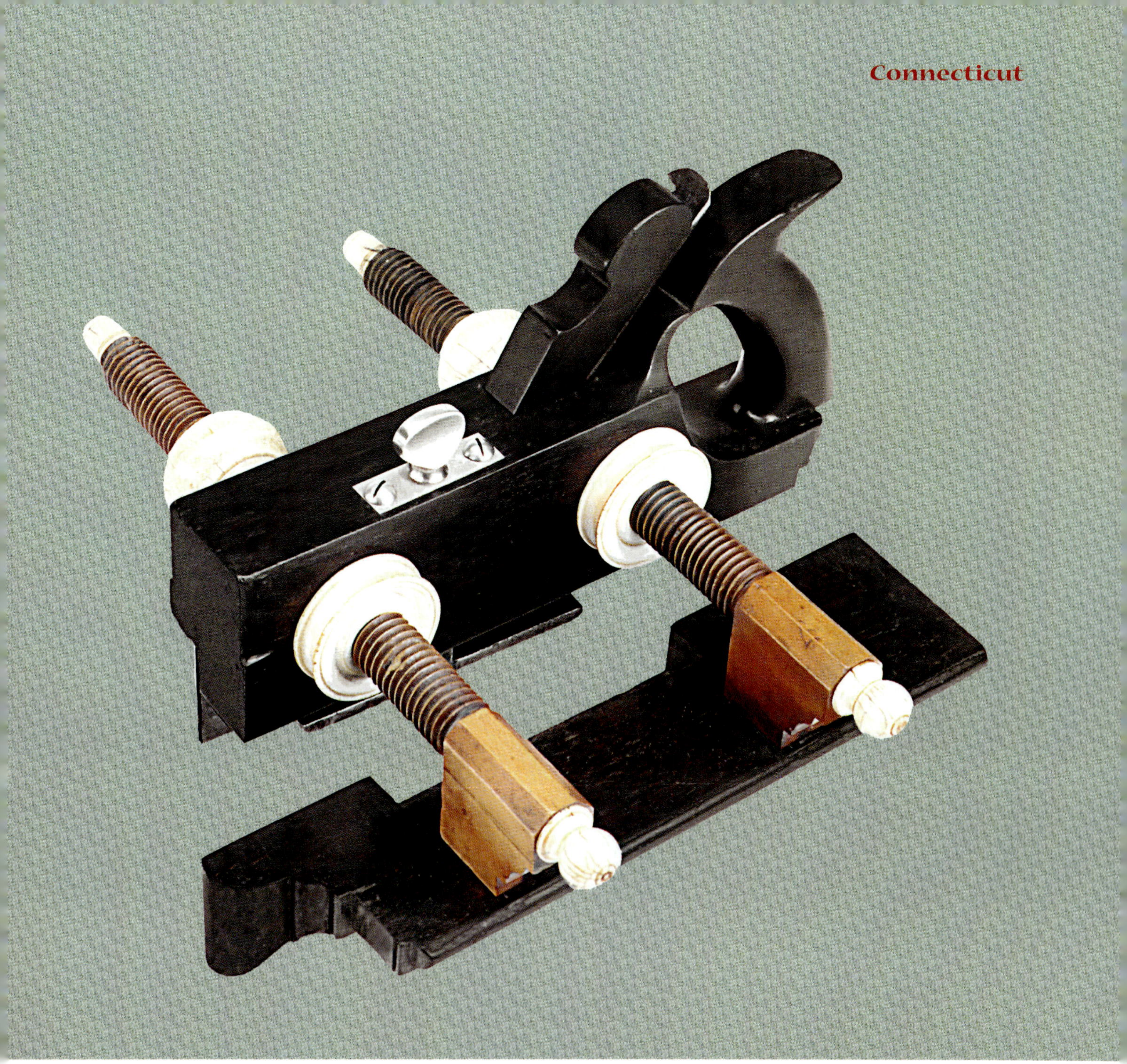

In spite of the addition of ivory nuts to this ebony plow, the change of the wood in the arms, from ebony to boxwood, detracts from the overall appearance of this plane when compared to the plane on the preceding pages. Ivory nuts are a very rare feature on plow planes, and this plane must have been very near the top of the line.

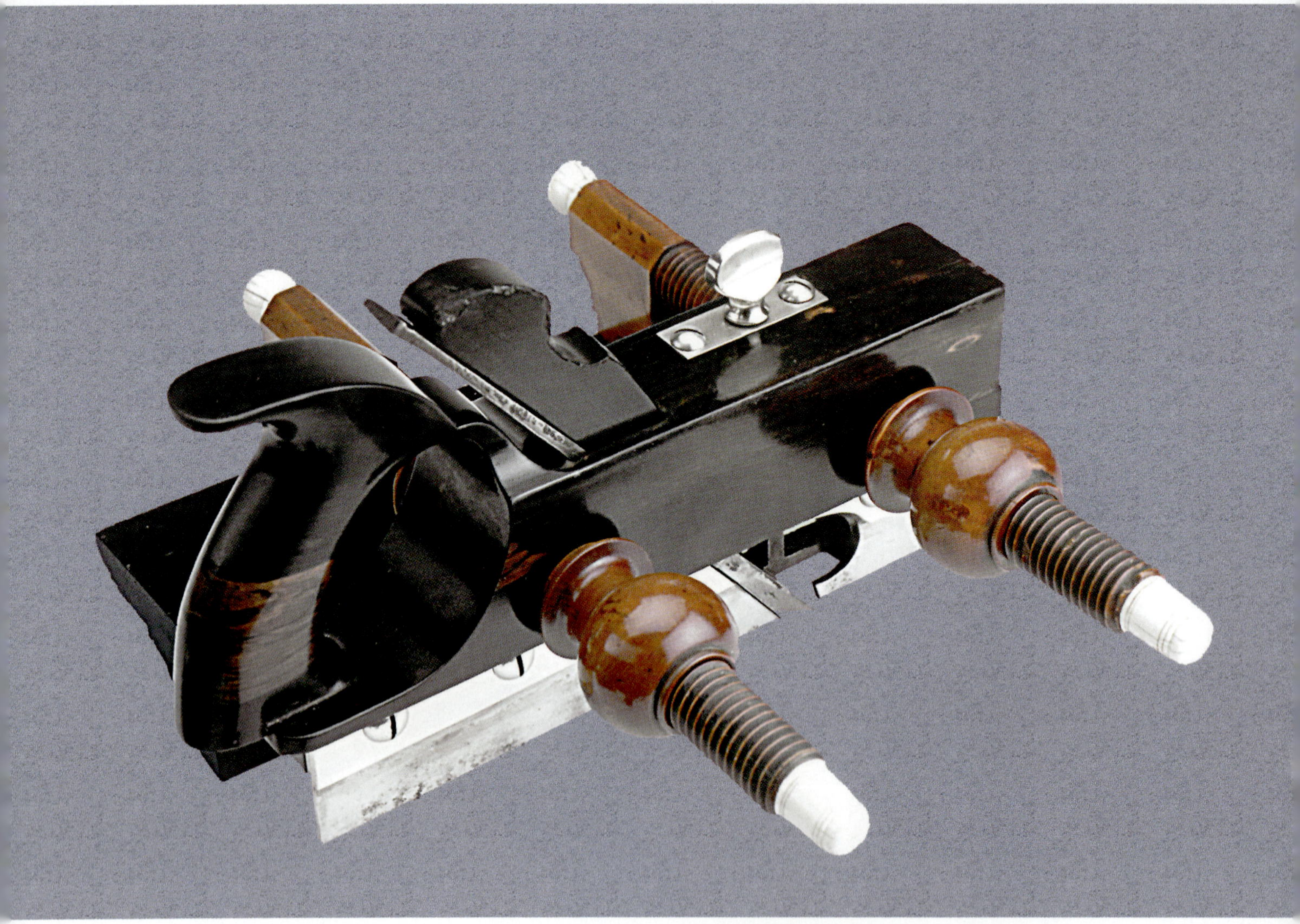

L. DeForest

Handled Macassar ebony screw-arm plow with boxwood arms and nuts, and silver mounting.

(From the collection of Don Rosebrook)

The many streaks of orange tan color mark this plane as being made from Macassar ebony. DeForest catalogs do not mention Macassar ebony but do mention zebra wood. From the planes observed to date, it seems likely that what we call Macassar ebony, they called zebra wood. The cost of articles made from zebra wood was identical to articles made from ebony, thus giving some credence to the idea that zebra wood was a form of ebony.

Macassar ebony, also known as Indian ebony, comes from the Celebes Islands and is somewhat more dense than African ebony. African ebony comes from southern Nigeria, Ghana, Cameroon and Zaire. Macassar may also have grayish brown streaks. It is related to ebonies from the Indian subcontinent and Sri Lanka.

L. DeForest

Handled mahogany screw-arm plow with boxwood arms, nuts and wedge.
(From the collection of Peter Hathaway)

L. DeForest

Unhandled Macasser ebony screw-arm plow with ebonized boxwood nuts.

(From the collection of Kenneth Thomas <<< Don Rosebrook)

G. W. Denison & Co.

Handled cocobolo screw-arm plow with boxwood arms and nuts. Silver mounted.

(From the collection of Kenneth Thomas <<<Don Rosebrook)

Cocobolo is a type of rosewood. Brazilian rosewood is the type that is used in most applications for rosewood. Cocobolo is about 20% heavier than Brazilian rosewood, much harder and very tough. It tends to contain more red and orange color than Brazilian rosewood.

G. W. Denison & Co.

Handled rosewood screw-arm plow with boxwood arms and nuts, and depth stop lock.

(From the collection of Don Rosebrook)

G. W. Denison, in partnership with Gideon Hull and Jedediah Harris, constituted G. W. Denison & Company. They operated in Winthrop, Connecticut from 1868 to 1884. The partners were all related by blood or by marriage. This company not only sold planes to hardware stores but also to other planemakers, including John Gladding, P. A. Gladwin, and Bodman & Hussey.

J. Denison

Unhandled rosewood screw-arm plow with boxwood arms and nuts. C mark **

(From the collection of Andrew Delans)

The light colored wood on the top surface of the plane body is the sapwood of the rosewood tree. The great contrast in color between rosewood heartwood and its sapwood has often been used to provide a pleasing look in tools and tool parts.

John Denison manufactured planes in Saybrook Connecticut. in partnership with his brother Lester from 1832-1840 as J. & L. Denison. John Denison continued in business by himself from 1840-1876. Lester continued to make planes in Winthrop, Connecticut from 1840 - 1866 and, upon his death, his adopted son, G. W. Denison continued his business under his own name and that of G. W. Denison & Co.

J. Denison

Unhandled ebony screw-arm plow with boxwood fence, arms and nuts, unusual shaped nuts, and interesting repair with brass strip on left side of body.   B mark *

(From the collection of Don Rosebrook)

The shape of the large nuts on this plane may be aesthetically pleasing but it is a highly unusual shape. Even more unusual is the repair on the left side of the body. The look of the repair is not offensive and it was apparently very effective because it has been on the plane for a long time and shows evidence of wear. The repair was for the side being split out by the wedge. This type of damage is often noted on plow planes.

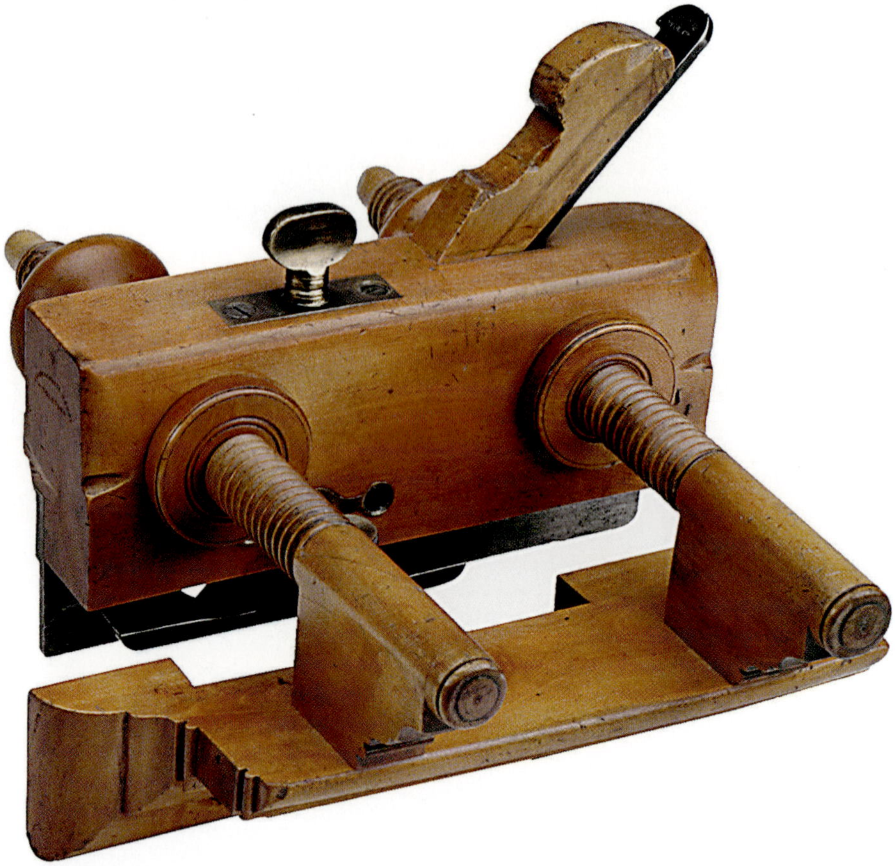

Arrowmammett

Unhandled boxwood screw-arm plane with depth stop lock. Made at the Arrowmammett Works of the Baldwin Tool Company.

(From the collection of Peter Hathaway)

The Baldwin Tool Company was founded in Middletown, Connecticut in 1836 and operated the Arrowmammett Works from then until 1857. It was located on the Arrowmammett River, hence the name given to some of the product line. In 1857, the works was sold to the Globe Manufacturing Company and it is unclear whether or not the Arrowmammett line of planes was continued.

James Bradley

Unhandled beech wedge-lock slide-arm plane with wooden depth stop and riveted fence and skate. Unplated skate.   A mark ****

(From the collection of Mickey Holmes)

James Bradley worked in New Haven, Connecticut around 1800.

Note that this plane has flat chamfers. The wooden wedge and use of rivets to hold the arms to the fence is typical of the period. The lack of brass plating on the skate should indicate an inexpensive plane.

C. Morehouse

Unhandled boxwood screw-arm plow. ****

(From the collection of Mickey Holmes)

Charles Morehouse was a planemaker in New Haven Connecticut from 1840 to 1846. There is thought to have been some potential interaction between Morehouse and W. H. Pond (also a New Haven planemaker) because Morehouse planes are known to exist carrying a Pond overstamp, and because some Pond planes have apparently used the same location die.

The shape of the flat nuts is distinctive but not unique. It will help identify planes by this maker.

Shelton & Osborn Mfg. Co.

Handled ebony screw-arm plow with boxwood arms and nuts. **

(From the collection of David Burge <<< Don Rosebrook)

Shelton & Osborn Mfg. Co.

Handled ebony screw-arm plow with boxwood arms and nuts. **

(From the collection of Dr. Ronald Pearson)

Note that the large nuts on this plane are neither shaped like the DeForest nuts, shown earlier, nor like the larger diameter Shelton & Osborn on the preceding page. It is not known if the large nuts are original to the plane.

Shelton & Osborn Mfg. Co.

Handled ebony screw-arm plow with boxwood arms and nuts, ivory tips, and silver mounting. **

(From the collection of Dr. Ronald Pearson)

The large boxwood nuts on this plane are very similar to the nuts on the first of the two preceding ebony plows. Note the difference between the nuts on the two ebony plows. This is unusual for the short working period of the company.

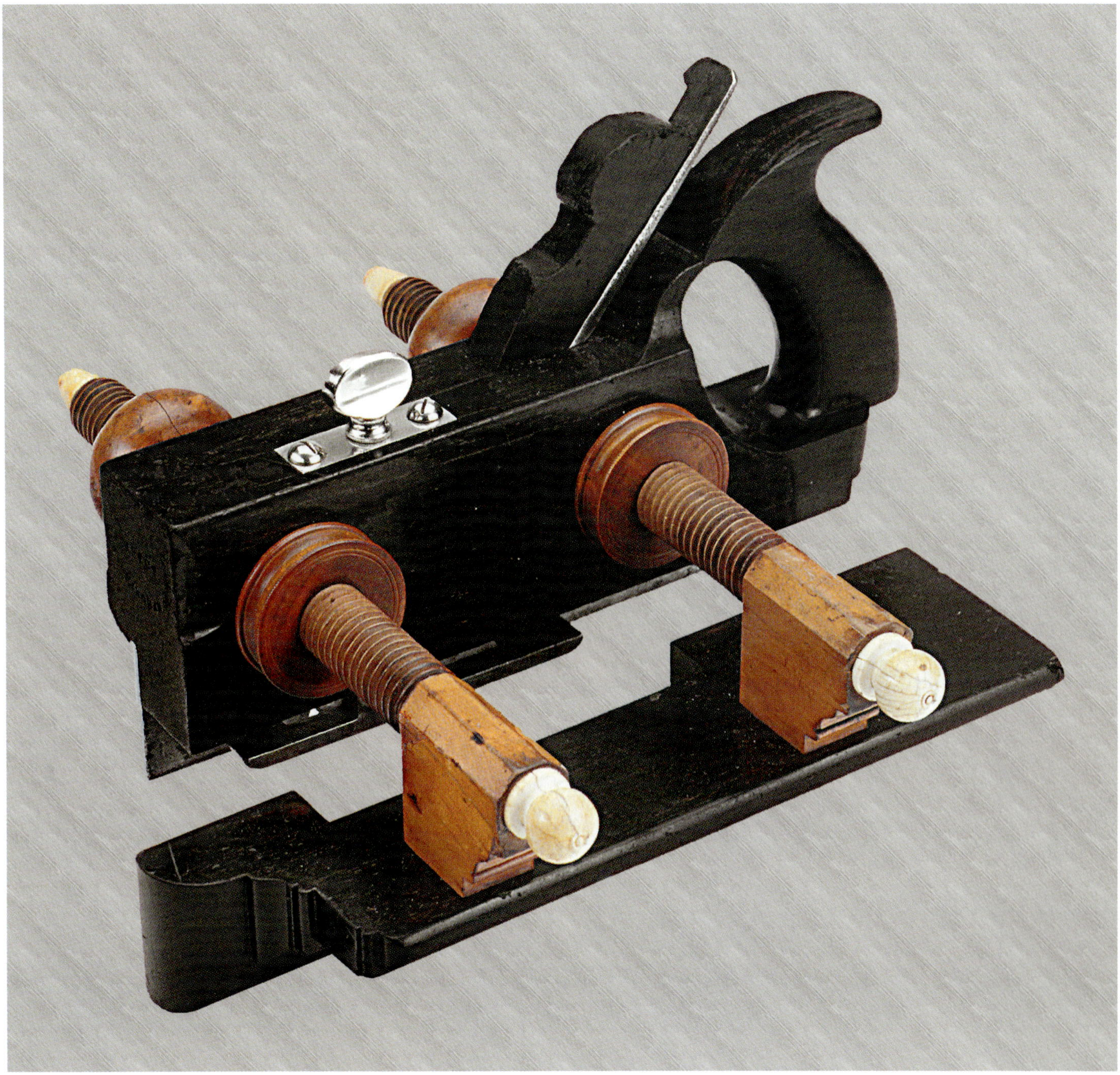

Shelton & Osborn is somewhat of a mystery. One school of thought is that their planes were made by DeForest, since both companies were located in Birmingham, Connecticut. Another school of thought is that Shelton & Osborn made the DeForest planes, in particular those which were not signed. It is interesting that the surviving plow planes from each company are from the high end of their production, often of ebony and, of these, often made with ivory and silver trim.

Shelton & Osborn Mfg. Co.

Handled ebony screw-arm plow with boxwood arms, ivory nuts and tips, and silver mounting. **

(From the collection of Dr. Ronald Pearson)

Note that the ivory tips and ivory nuts on these Shelton & Osborn plows are a close match for the ivory trim on some of the DeForest plows. For example, the tips on the Shelton & Osborn plows match the ivory tips on the DeForest ebony plow with ivory nuts, but not on the other two DeForest ivory trimmed plows.

E. & B. Spencer

Unhandled beech wedge-locked slide arm plow, interesting wedge-lock slanted style depth stop, and riveted arms. **

(From the collection of Mickey Holmes)

E. & B. Spencer used the signature "Spencer" and made planes in New Haven, Connecticut. B. Spencer was probably Benjamin Spencer, who lived in Saybrook Connecticut from 1810 to 1830 and was later listed as a planemaker in New Haven. Pollak does not report any identification of E. Spencer. The estimated working period of Spencer was 1810 - 1850.

The wedged wooden depth stop is novel in the authors' experience. It eliminates the need for a depth stop lock.

C. Cole

Unhandled birch screw-lock, slide-arm plow with bone depth stop and 2 bone wear plates on the fence.

(From the collection of Robert Wheeler)

Pollak has no location for this maker but speculates that he was making planes around 1790 or 1800.

J. Gallup

Unhandled fruitwood screw-lock, slide-arm plow with depth stop lock. Very narrow-topped thumb screws and mortises in body to receive arm shoulders.

(From the collection of Mickey Holmes)

Pollak speculates that this maker was working in New England around 1790.

This plane has mortises in the right hand side of the body to receive the fence, instead of a mortise in the fence to receive the depth stop. Either way allows the fence to get close to the skate.

H. Wheeler

Unhandled birch screw-lock, slide-arm plow with depth stop lock. Date 9-11-1809 stamped in nose.

(From the collection of Mickey Holmes)

Little is know about this maker except for the presence of the plane with the 1809 date inscribed on it.

Because this plane was little used, the planemaker's penciled lines used to lay out the various cuts are still clear. (The color has been lightened and slightly altered to better show the pencil lines.) This plane presents an excellent example of how planemakers began their work.

R. Ionson

Unhandled beech screw-lock, slide-arm plow with depth stop lock and ebony wear strips in top of arms. C mark **

(From the collection of Mickey Holmes)

This is probably an early spelling of Jonson. Pollak speculates that Jonson worked in the lower Vermont/New Hampshire section of the Connecticut River Valley. The working period is believed to be 1800 to 1810.

Note the ebony wear strips in the top of the arms to protect the arms from the pressure exerted by the thumb screws used to lock the fence in place. There are many examples of worn arms pictured in this book.

2 New York & New Jersey Planemakers

While there were planemakers scattered throughout New York, from Buffalo and Rochester to Albany and New York City, it is the products of the New York State Prisons that are best known. The contract for prison labor was given to planemakers for approximately 50 years, from about 1825–1875. The output of the prison labor was very high and much of it was sold to wholesalers and hardware dealers for them to resell under their own names.

The planes in this section begin with the earliest contractor, T. J. McMaster, and progress in chronological order to the Auburn Tool Co. The reader can follow the flow of changes in style and the eventual evolution of new styles for the succeeding contractors. The brother of T. J. McMaster, Z. J. (Zalmon) McMaster, is credited with one of the most unusual American plow planes. The plane, thought to be a prototype, features a single arm capped with an ivory ball containing a gearing arrangement that moves the fence in and out.

There was considerable use of ivory for decorating prison-made plow planes. In addition, a large number of rosewood and ebony plow planes are observed. Perhaps the low cost of prison labor made it possible for the contractors to absorb the cost of rare woods such as rosewood and ebony and produce a proportionately larger number of these planes.

In addition to many examples of prison-made planes, a short feature is presented covering the prison, the contractors, and one of the men who worked there.

The presentation of prison-made planes is followed by examples of work from Rochester, Albany and New York City planemakers. The New York planes presentation concludes with a wooden plow plane that was never meant to be used — a wooden model of a metallic plow plane — by a maker from New York City who is usually associated with Pennsylvania, because most of his metallic planes were made in Wilkes Barre.

Only a few New Jersey plow planes have been presented, and some of the selected planes show the influence of Israel White, a Philadelphia planemaker who is covered in the next section on Maryland and Pennsylvania planes.

Plane Making In The Auburn New York State Prison

Plane making in the New York State Prison system was big business, particularly in the prison at Auburn, N. Y. The enterprise started in 1823 by Truman J. McMaster and Nathaniel Garrow with the production of both carpenter's planes and plane irons. These gentlemen held the contract until 1833, when Alonzo McMaster and Jacob Young won the contract and continued the business under the name of Young & McMaster until 1838. In 1838, the contract was won by the firm of Z. J. McMaster & Co. which was composed of Z. J. McMaster, Paul D. Cornell and Aurelius Wheeler. In 1847, the contract was taken over by a firm called Casey, Kitchell & Co. and composed of George Casey, Adam Miller, Josiah Douglass and Nelson Kitchell. George Casey bought the interests of his partners in 1857 and operated under the name of Casey & Co. In 1858, he formed a new company with J. N. Starin, Nelson Fitch, Abijah Fitch, Noah P. Clark and Alonzo Beardsley. The new company was called Casey, Clark & Co. until October 8th, 1864, when a stock company was formed, incorporated with a capital of $75,000, and called the Auburn Tool Co. George Casey was elected president.[1] In 1869, Augustus Howland formed a company with Nelson Fitch, C. P. Fitch, Edward Myderse, and Charles Tuttle and took the contract for prison labor away from Auburn Tool Co. Auburn Tool Co. regained the contract in 1874 and held it until August 1, 1877, at which time it moved off from the prison grounds.

The 1865 census reported that the Auburn Tool Co. utilized 35 tons of iron, 8 tons of steel and 30,000 board feet of lumber; employed 50 men and steam power; produced 30,000 pairs of skates, 35,000 planes and 25,000 plane irons. The average monthly wage of the men was $16 (prison labor rates). The value of the production was $87,000.

The next page contains a view of the Auburn State Prison as of 1869.[2] The caption reveals that the prison was an industrial center containing a sash & blind shop, tool shop, hame shop, shoe shop, cabinet shop, reaper & mover shop, axle shop and foundry. Presumably the planes were made in the tool shop.

George Casey

George Casey was born in Dover, N. Y. in March 1807 and served an apprenticeship as a carpenter. He began his career as a builder in Auburn and became a major player in that business, having built a city block of business buildings, a number of homes, St. Peter's Church, the Universalist Church, factory buildings, the Casey Opera House, and the buildings of the State Fair held in Auburn in the fall of 1846. He then turned his talents toward the manufacture of carpenters' and cabinet makers' tools and is said to have greatly improved the planemaking machinery of the prison factory. His principal invention in this regard was a machine for cutting the throats in planes. George Casey was a major force in plane making with prison labor from 1847 to 1877 and thereafter, with Auburn Tool Co. until his death in November 1890.[3]

William Embody

William Embody was born on July 11, 1849 in Herkimer County, N. Y. He was first listed in the Auburn city directories as a plane maker from 1867-1868. During 1875-1877, he was listed as Foreman of the Tool Shop in Auburn State Prison. When Auburn Tool Co. left the prison in 1877, he continued to be listed as a tool maker and plane maker. He died on April 26, 1906 at age 57 in Aurelius, N. Y. (near Auburn).

---

[1] The foregoing information was taken from Elliot G. Storke, *History of Cayuga County, New York*, Heart of the Lakes Publishing Company, Interlaken, New York, 1980, page 170.

[2] The Annual Report of the Inspectors of State Prisons for the year ending Sept. 30, 1869, Argus Co. Printers.

[3] The foregoing information was taken from George Casey's Obituary in the Auburn Bulletin of Wed. Nov. 2, 1890 entitled "DEATH OF GEORGE CASEY Another Old and Useful Resident Passes to Final Rest."

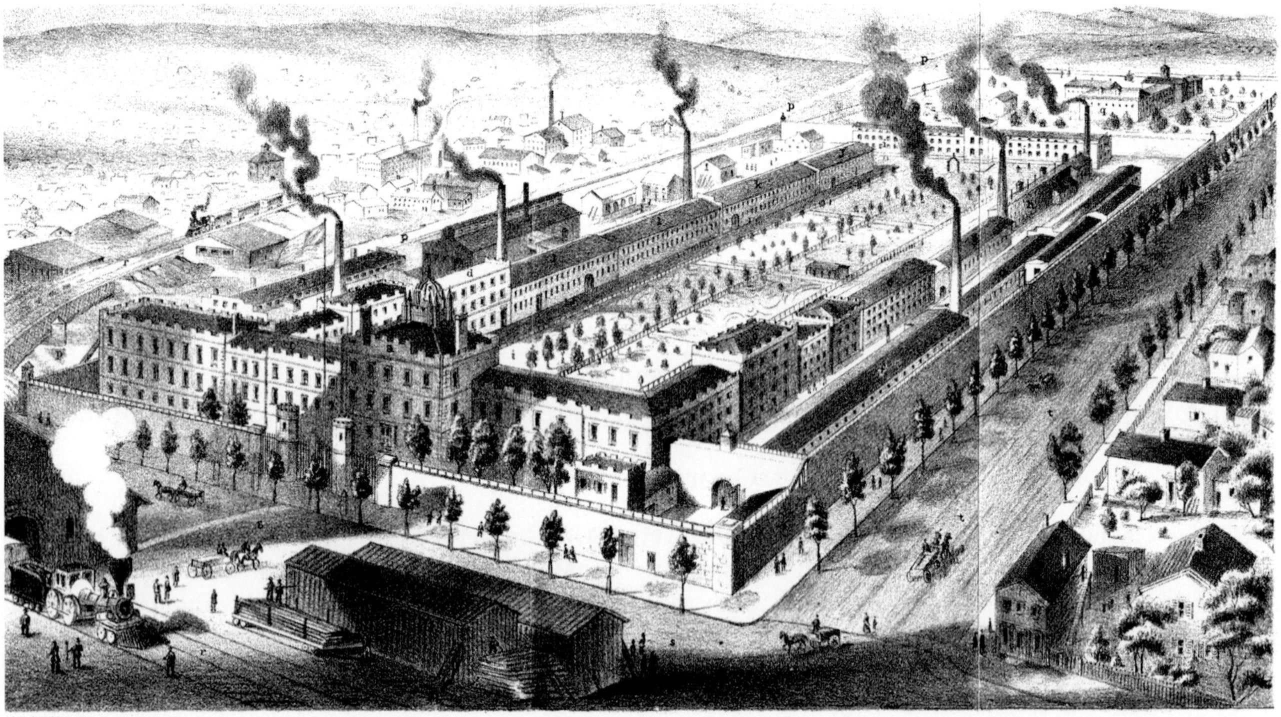

AUBURN STATE PRISON, 1869.

STATISTICS, 1869.—Location, in the city of Auburn, an outlet of Owasco Lake, fronting east. Established in 1817. Area of Grounds, 400 by 100 feet. Front wall 20 feet high. Side and rear walls 30 feet high. Center main building 56 feet high, wings 45 feet high. Front-length of building 387 feet. Valuations,—Real, $542,795.00; Personal, $63,516.43; Total, $706,211.43. Average number of convicts for the year, 949 11-12.

REFERENCES.—A, Agent and Warden's dwelling, and offices of Clerk, Agent and Warden and Inspectors. B B B B, North wing, containing 550 Cells. C C, Dining Hall in 1st story and Kitchen in rear,—2d story, Chapel, Chaplain's Office and Library,—3d story, Hospital and Store Rooms. D D D, South extension containing 442 Cells. E, New extension (unfinished), containing 300 Cells. (Total Cells, when finished, 1292.) F, Sash and Blind Shop. G, Tool Shop. H, Hame Shop. I I, Shoe Shop. J, State Shop in 1st story; Shoe Shop in 2d story. K, Cabinet Shop. L L, Reaper and Mower Shop. M, Foundry. N, Axle Shop. O O O O, Vacant Shops, P P P, Owasco outlet. Q, Asylum for insane Convicts R R. N. Y. C. Railroad. S S. State Street. T T, Wall street,— 2 "Copper John."

William Embody's Planemaking Tools

Andrew Delans obtained many of William Embody's planemaking tools and some other memorabilia which are shown below and on the following pages.

Saws, Knives and a Rasp

Planemakers Floats

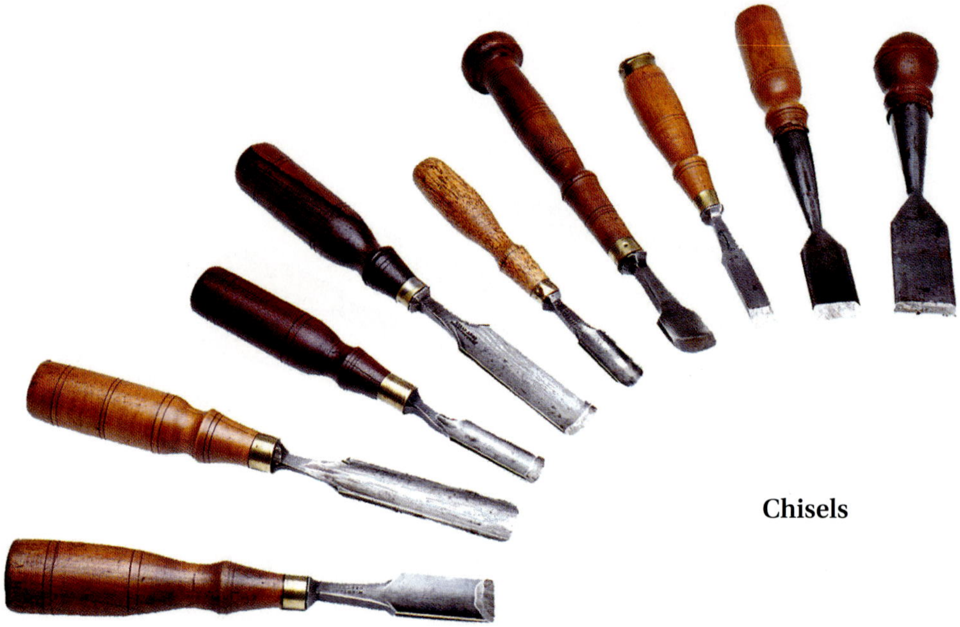

Chisels

Jack Plane, Cheeking Chisels and a Pattern Piece for Throats

Prison-Made Inlayed Box
This box was made in Auburn prison, probably in the middle 1870s.

T. J. McMaster & Co.

Unhandled beech screw-arm plow with boxwood arms and nuts. D1 mark *

(From the collection of Don Rosebrook)

Truman McMaster was a brother of Z. J. McMaster. There is much confusion in our reference material about the McMaster brothers, and the various prison labor contracts, both within Pollak and between Pollak and Storke. Pollak claims that Truman McMaster made planes with convict labor in Auburn Prison from 1825-29, and at Sing Sing Prison from 1829-1838. In 1839, he is said by Pollak to have defaulted on his contract, and it was taken up by his brother (Alonzo) and a partner who made planes as Young & McMaster. However, Young & McMaster do not have a Sing Sing mark. Pollak also claims that Dunham & T. J. McMaster made planes in Auburn Prison from 1821-1825. To add to the confusion, Z. J. McMaster was said to have held the Sing Sing contract from 1839-1843 and the Auburn contract from 1846-1847 and Z. J. does have a Sing Sing mark. Storke, who is a writer more nearly contemporary with the operating times than Pollak, has a very different story, as related in the introduction to the Auburn Prison feature.

Z. J. McMaster & Co.

Unhandled ebony screw-arm plow with single pentagonal arm with ivory ball at the outboard end. Arm carrying a threaded rod underneath is fixed to the fence by a massive arch.

(From the collection of Don Rosebrook)

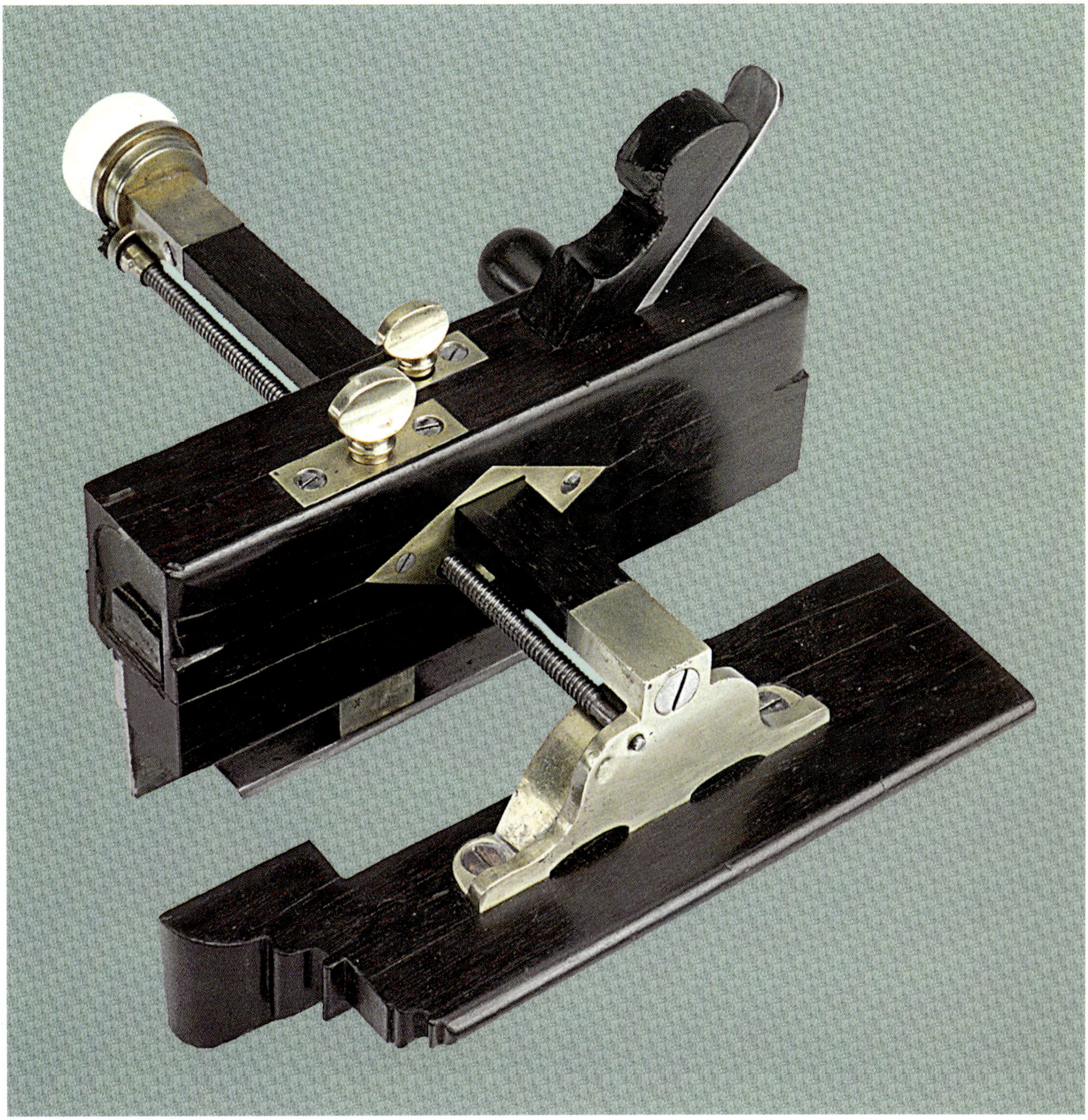

This one of a kind plane dates to the time of the McMaster contract in Auburn. The ivory ball contains a gear, which, when the ball is turned, drives a gear on the rod beneath the arm. The threaded rod passes through a threaded piece in the body. Thus, turning the rod will move the fence. The rod is visible in the top center picture which looks at the arm from below. The plane is extremely heavy due, in large part, to the large amount of brass on the plane. No other signed, single arm American plow is known. In addition, the authors are not aware of any other American plows that have the maker's mark in a church window.

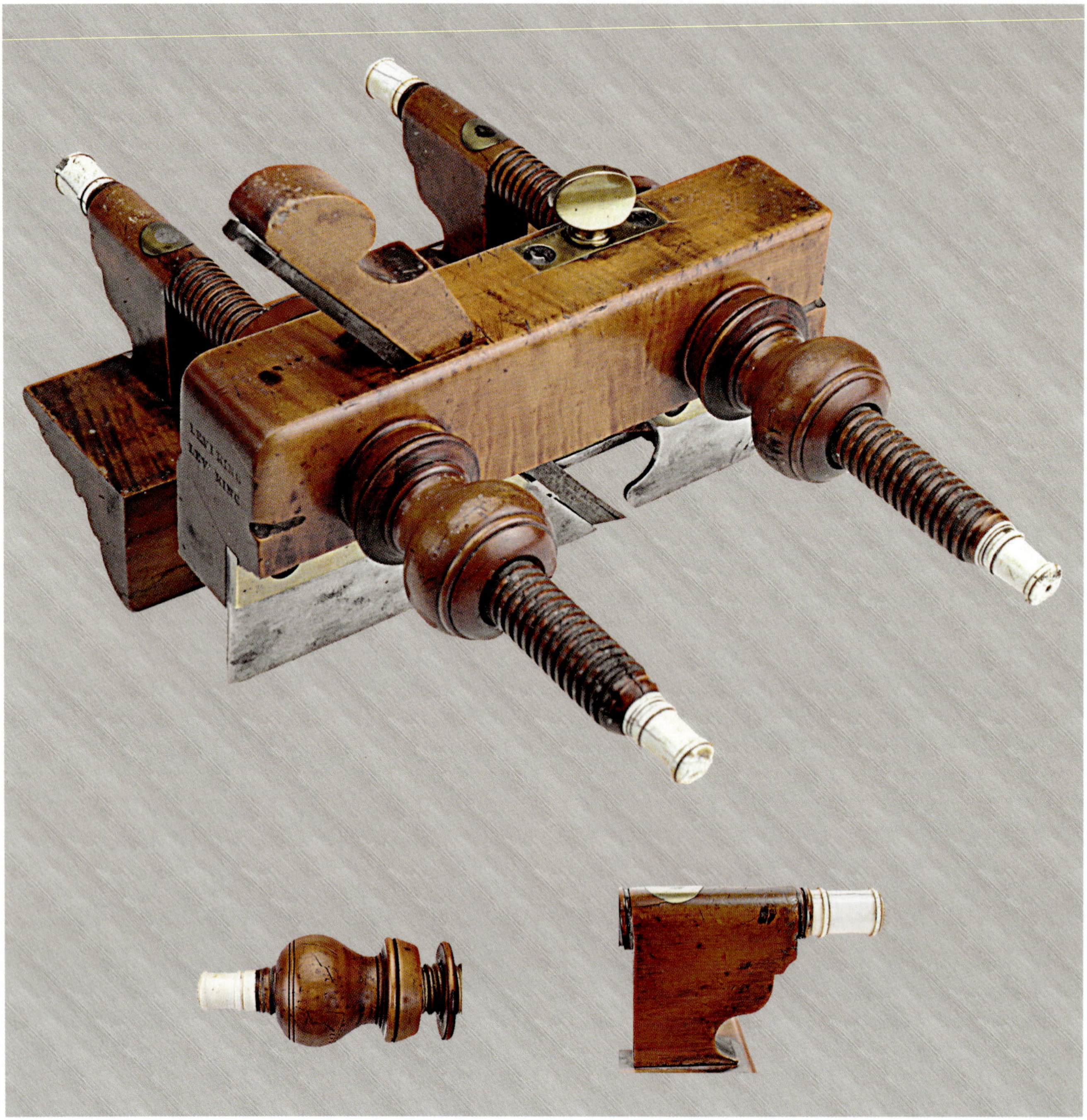

Z. J. McMaster & Co.

Unhandled boxwood screw-arm plow with ivory tips and boxwood washers.

(From the collection of Don Rosebrook)

The molding shape on the shoulders of the arms is among the most distinctive found on manufactured planes. The ivory tips here are certainly the inspiration for the tips on early Casey, Kitchell & Co. planes.

Zalmon McMaster was a brother of Truman and Alonzo McMaster, and is said to have been making planes at Sing Sing Prison from 1839-1843. This conflicts with information reported elsewhere in Pollak, which states that Alonzo McMaster was the partner in Young & McMaster. Perhaps there were two simultaneous contracts for prison labor. Z. J. McMaster did make planes at Auburn Prison from 1839-1843 and 1846-1847. No information has been reported for Z. J.'s activities between these three periods.

Z. J. McMaster & Co.

Unhandled beech wedge-lock slide-arm plow with boxwood wedges, and brass collars on the ends of the arms. I mark ***

(From the collection of Frank Kosmerl)

Z. J. McMaster & Co.

Unhandled beech screw-arm plow with boxwood arms and nuts. C mark **

(From the collection of William Curtis)

A. Miller

Unhandled boxwood screw-arm plow with ivory tips and boxwood washers. ***

(From the collection of Don Rosebrook >>> ???)

A. Miller is reported to have operated somewhere in N. Y. in approximately 1850. The appearance of this plane would indicate that A. Miller was probably a hardware dealer and that he was in business during the 1840s.

When the appearance of this plane is compared with that of Z. J. McMaster's unhandled boxwood with ivory tips on page 148, the origin of this plane becomes apparent. If McMaster did not make the plane, it is at least a good copy, with only a slight difference in the flat nuts.

Casey, Kitchell & Co.

Handled boxwood screw-arm plow with boxwood washers, double ended fence, and ivory tips. *

(From the collection of Andrew Delans)

This company bought the contract for prison labor from Z. J. McMaster & Co. and made planes in Auburn prison from 1847 to 1856. In 1856 or 1857, George Casey bought out his partners and operated under his own name for in 1857. The folds in the back of the handle, where it meets the body, seem to be peculiar to this maker. The hardware store price of $9 is still on the plane (see handle detail above).

This plane illustrates the shape of the early ivory tips on Casey, Kitchell & Co. planes. The ivory tips on this plane clearly show the McMaster influence, although they are not as long as the McMaster tips. Casey, Kitchell & Co also continued the use of washers behind the large nuts.

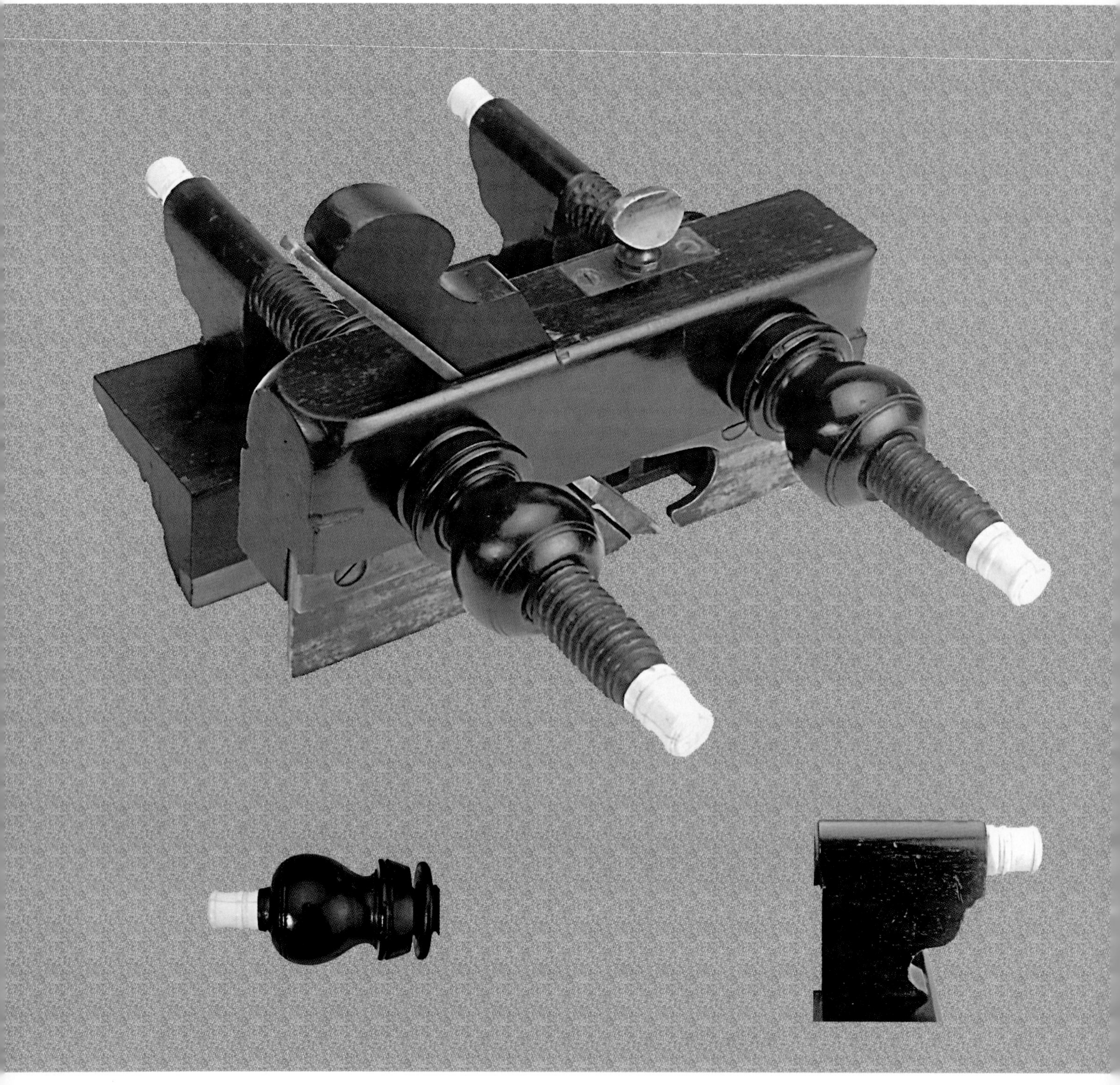

Casey, Kitchell & Co.

Unhandled ebony screw-arm plow with ebony washers, and boxed fence. *

(From the collection of Andrew Delans)

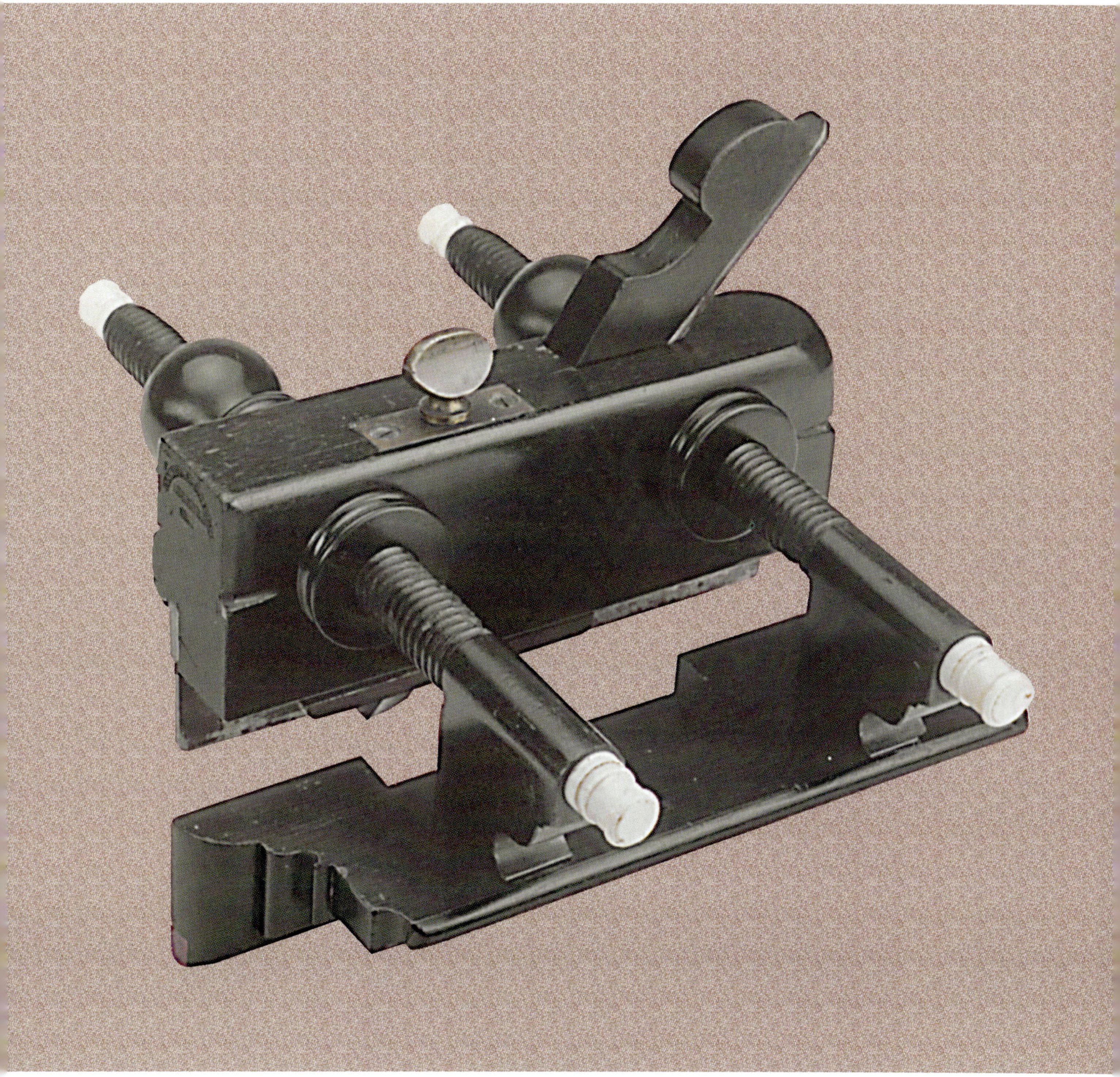

While most features of Casey, Kitchell & Co. planes were similar in handled and unhandled planes, the double-ended fence was used only on the handled planes.

Casey, Kitchell & Co.

Handled boxwood screw-arm plow with boxwood washers, double ended fence, and ivory tips. The depth stop is apparently an aftermarket addition. Screws have been brass plated. *

(From the collection of Frank Kosmerl)

Note that the shape of the ivory tips has changed and is moving toward the shape of the tips on Casey & Co. planes.

Casey, Kitchell & Co.

Handled ebony screw-arm plow with boxwood arms, nuts and washers, double ended fence, and ivory tips. *

(From the collection of Don Rosebrook)

The ivory tips on this plane represent the late stages of Casey Kitchell design. They need only to flare out again at the tips to have the shape used by Casey & Co.

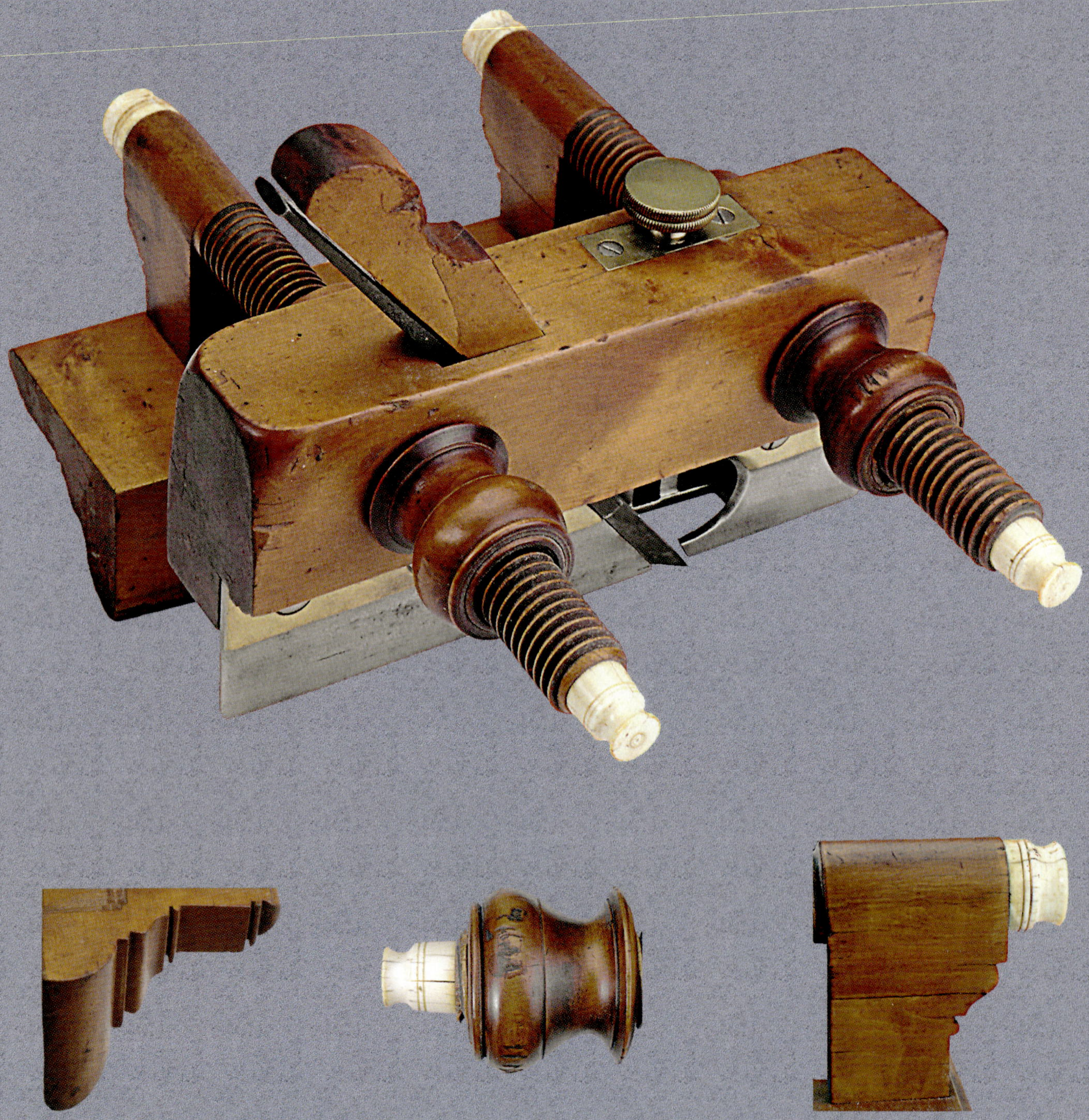

Casey & Co.

Unhandled boxwood screw-arm plow with ivory tips and depth stop lock.

(From the collection of Andrew Delans)

George Casey made planes in Auburn, New York under this name in 1857, using prison labor. Casey & Co. was the successor to Casey, Kitchell & Co. George Casey was involved in a number of other partnerships that made planes in Auburn using prison labor.

Notice the distinctive shape of the ivory tips.

Casey, Clark & Co.

Handled rosewood screw-arm plow with boxwood arms, ivory nuts and ivory tips, and depth stop lock.

(From the collection of Andrew Delans)

This company, the successor to Casey & Co., made planes in Auburn New York using prison labor from 1858-1864.

Casey, Clark & Co., like other George Casey companies, was a large operation and, in 1860, employed 65 hands, making $45,000 worth of planes and plane irons.

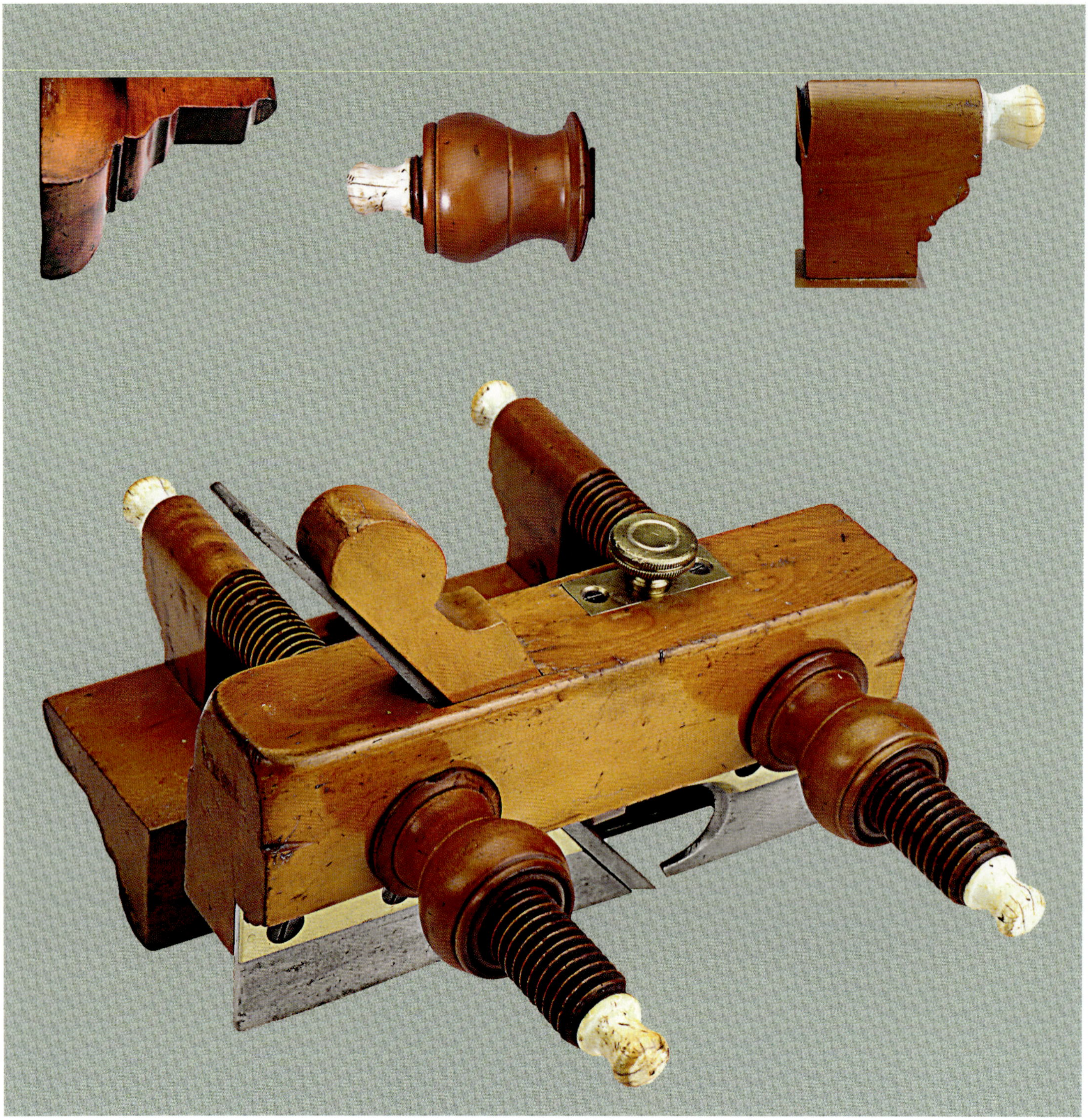

Casey, Clark & Co.

Unhandled boxwood screw-arm plow with ivory tips and depth stop lock.

(From the collection of Andrew Delans)

Note that the shape of the ivory tips has changed and the tips are substantially smaller than ivory tips on planes of the predecessor companies.

Casey, Clark & Co.

Unsigned, handled rosewood screw-arm plow with boxwood nuts, ivory tips, and a depth stock lock.

(From the collection of Don Rosebrook)

Note the unusual donut shaped flat nuts. These nuts seem to be an exaggerated version of the flat nuts on the preceding plane. Several companies changed the style of their nuts during the years in which they operated, and some of them, such as Greenfield Tool Company, had major changes during their operating lifetime. However, few of them changed the flat nuts.

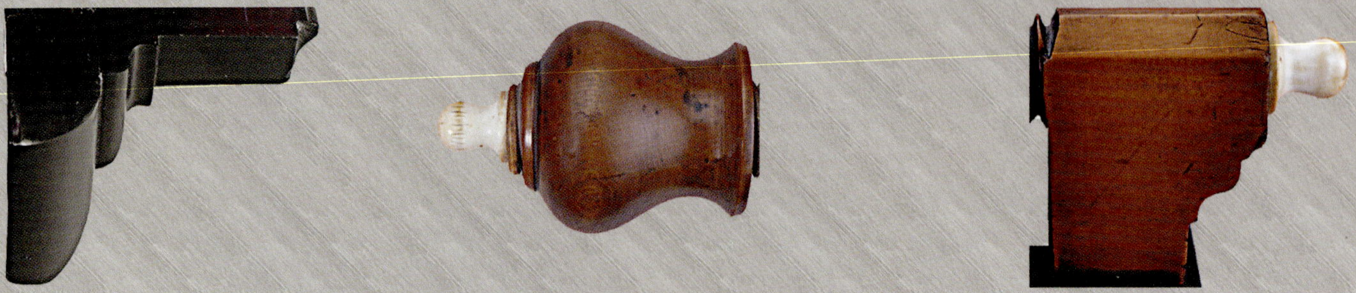

A. Howland & Co.

Handled ebony screw-arm plow with boxwood arms and nuts, ivory tips, and a depth stop lock.

(From the collection of Andrew Delans)

Augustus Howland was a bank president who formed a company to use Auburn Prison inmates to make planes (and possibly other tools). Howland won the contract for prison labor held by the Auburn Tool Co. They made planes under this contract from 1869-1874. Company testimony before the N.Y. State Commission on Prison Labor in1870 claimed that the hardware trade did not consider the quality of prison planes equal to that on the outside.

The ivory tips on this plane would seem to reflect the position in time of the Howland contract for prison labor. Planes made by the Auburn Tool Co. during period immediately preceding Howland's contract were fastened on the arms with screws. The later Auburn tips (from the subsequent contract) were somewhat larger versions of the Howland tips.

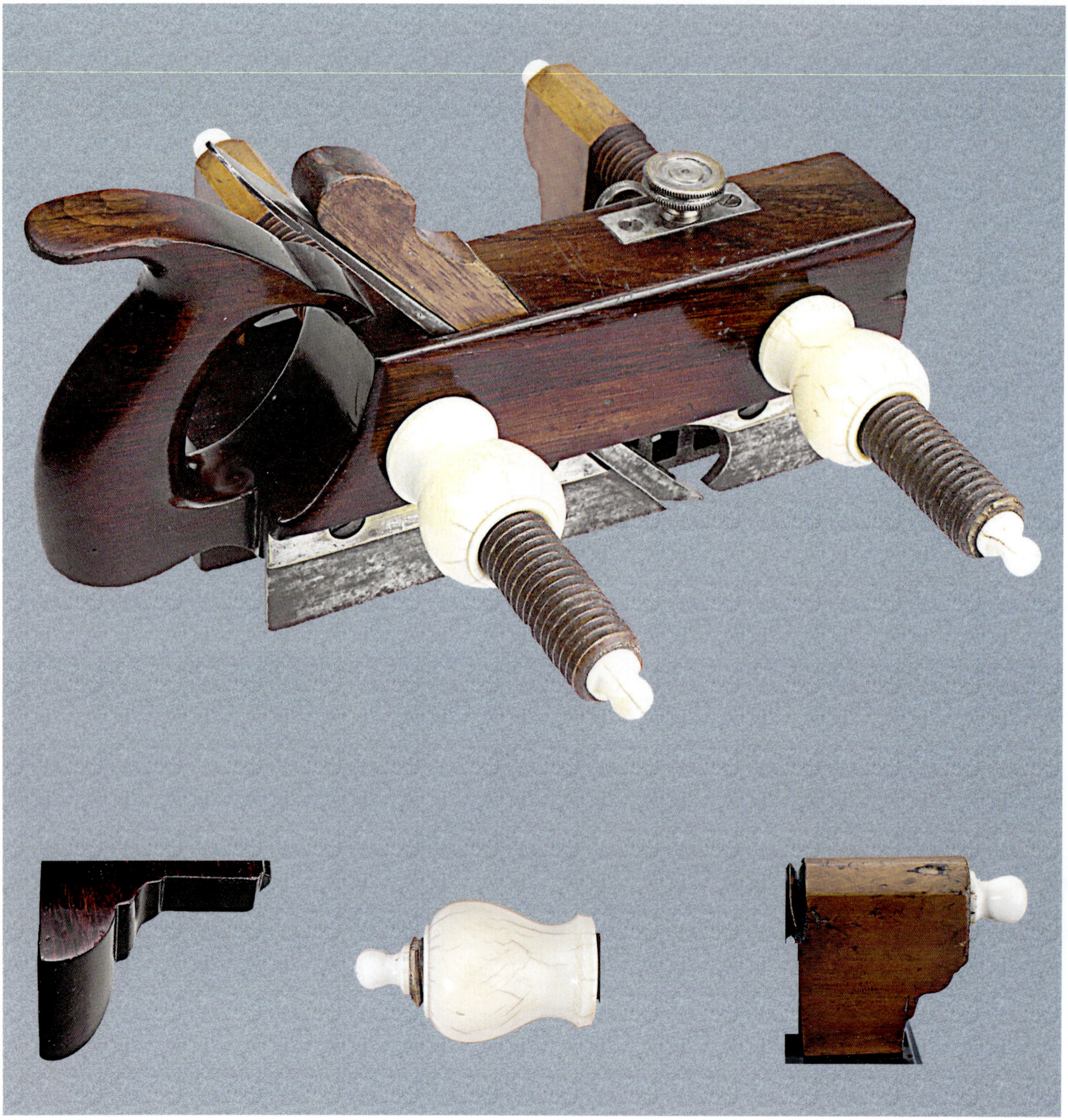

A. Howland & Co.

Unsigned, handled rosewood screw-arm plow with boxwood arms and wedge, ivory nuts and ivory tips, depth stop lock and silver mounting.

(From the collection of Andrew Delans)

The ivory nuts on this plane are essentially identical to those on Auburn planes. The ivory tips, while retaining the same general shape, have become still smaller.

A. Howland & Co.

Handled ebony screw-arm plow with boxwood arms and nuts, ivory tips and a depth stop lock.

(From the collection of Thomas Witte >>> ???)

This Howland plane was discovered in the shipping box as shown above. This is the only such box known to have been discovered to date. The shipping box included the 8 blades shown at the top of the page. They range in width from 1/8" to 5/8". Plane blades were made using a piece of good steel, which could form a good edge, that was forge welded to a softer piece, as is shown on the center picture above.

Auburn Tool Co.

Handled boxwood screw-arm plow with ivory nuts and 2 ivory tips, and a depth stop lock.

(From the collection of Andrew Delans)

The Auburn Tool Co. used convict labor in the Auburn prison to produce both planes and plane irons. It had the contract for prison labor from 1864-1865 and 1874-1877. Auburn Tool Co. was the successor to Casey, Clark & Co., and George Casey was its president. It existed from 1864 to 1893 at which time it merged with the Ohio Tool Co.

The use of only two ivory tips, even on expensive planes like this one, was common on planes made by the Auburn Tool Company. It was, however, occasionally used elsewhere.

Auburn Tool Co.

Unhandled rosewood screw-arm plow with boxwood arms and nuts, 2 ivory tips, and depth stop lock.

(From the collection of Frank Kosmerl)

The ivory tips are held on with a screw through the end. These are apparently the earlier form of the tips.

Note the shape of the ends of the arms and of the ivory tips and compare them to the ivory tips on the plane on the next page. Those on the next plane are more slender and elongated.

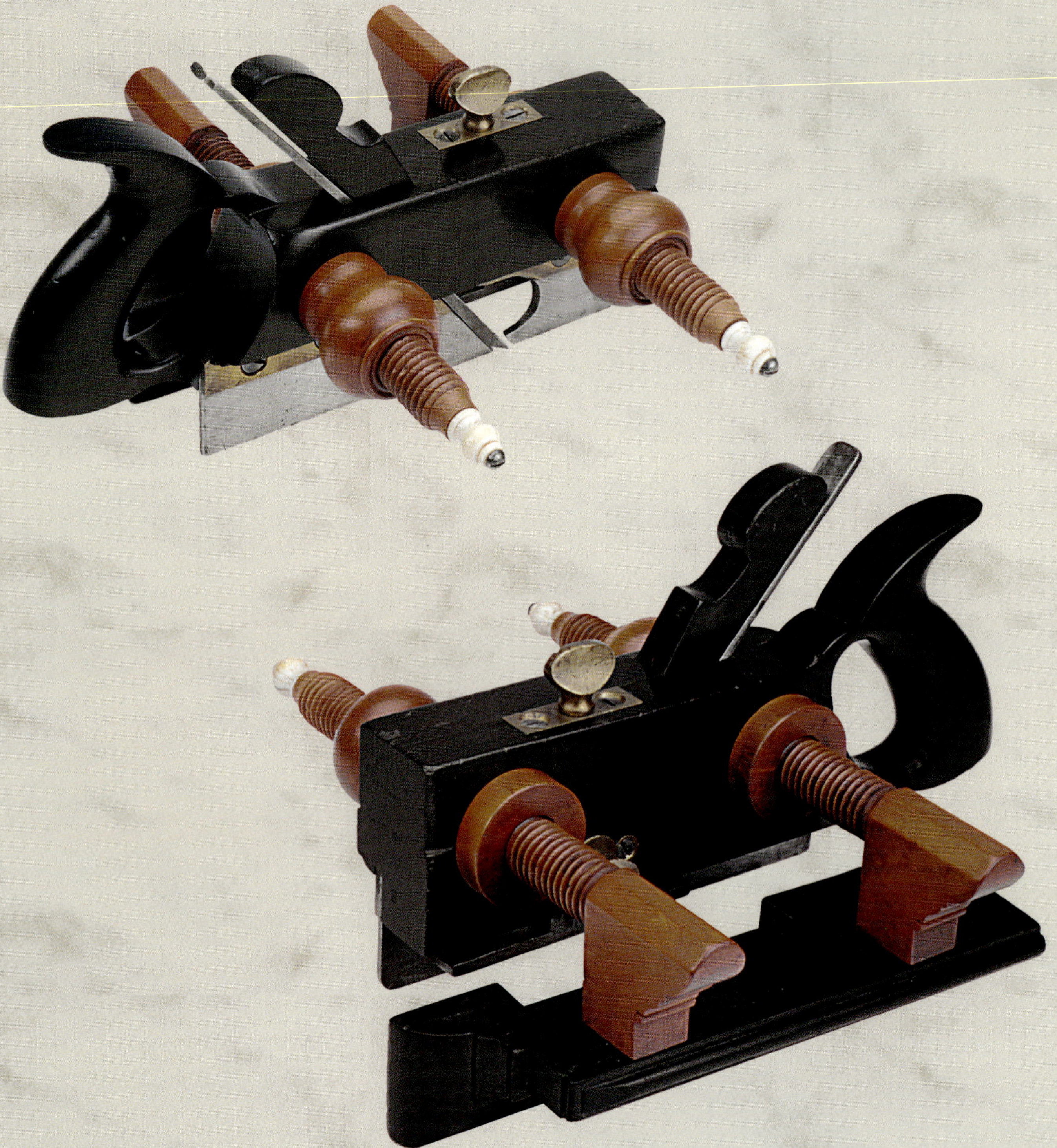

Auburn Tool Co.

Handled ebony screw-arm plow with boxwood arms and nuts, two ivory tips, and a depth stop lock.

(From the collection of Peter Hathaway)

D. R. Barton

Handled beech screw-arm plow with boxwood arms and nuts.

(From the collection of Frank Kosmerl)

David R. Barton founded this company in 1832 in Rochester, New York. Barton was involved in a number of partnerships and plane making was not necessarily the major product of his companies.

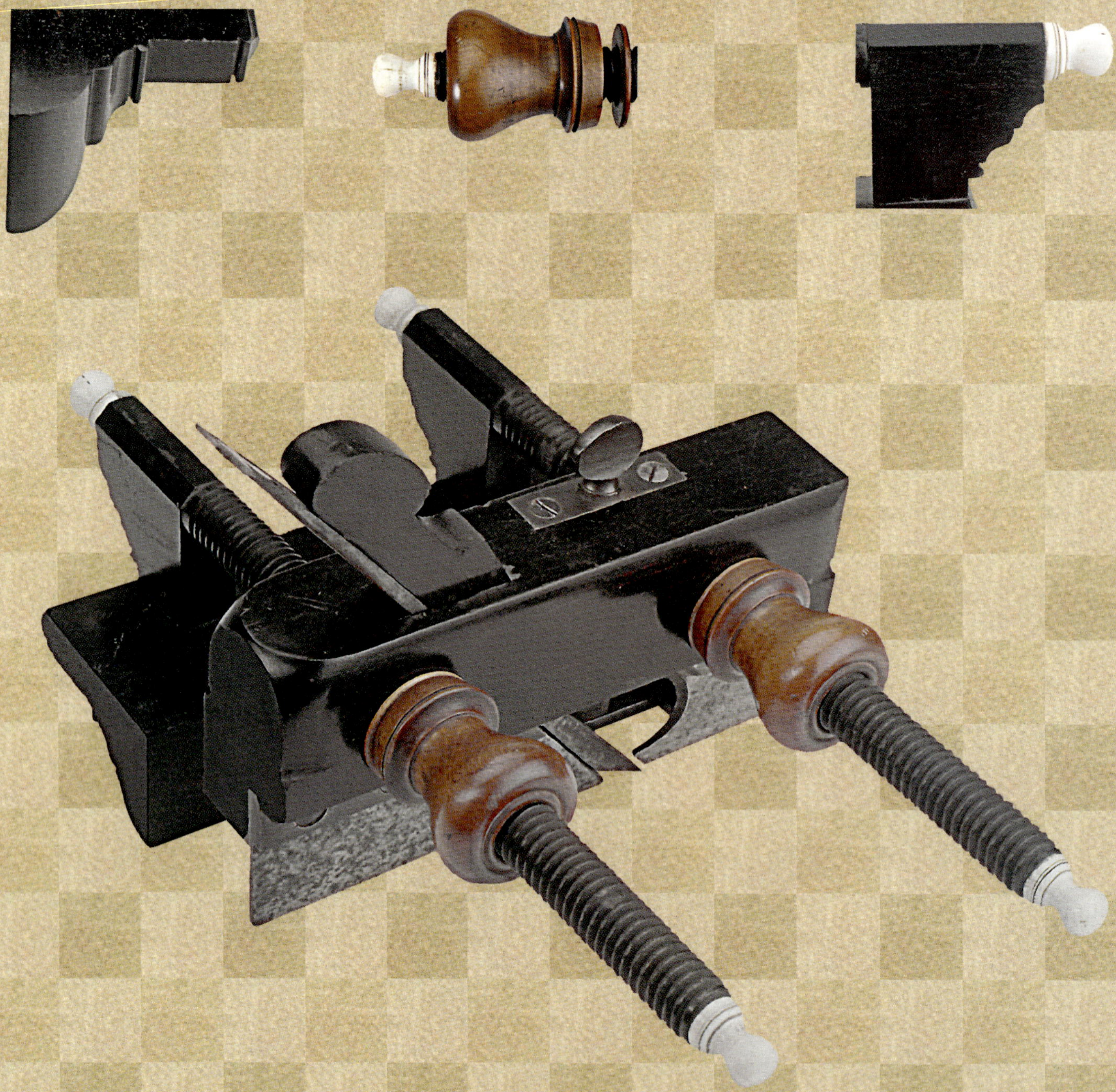

D. R. Barton

Unhandled ebony screw-arm plow plane with boxwood nuts and washers, and ivory tips.

(From the collection of Frank Kosmerl)

Barton companies are well known for edge tools. From 1855-1861 they utilized prison labor in Kingston, Ontario. Barton offered wooden planes until 1923. He obtained some of his planes from the Greenfield Tool Co. (see the planes on pages 9 and 12.)

Barton was one of only a few makers to use washers with the large nuts.

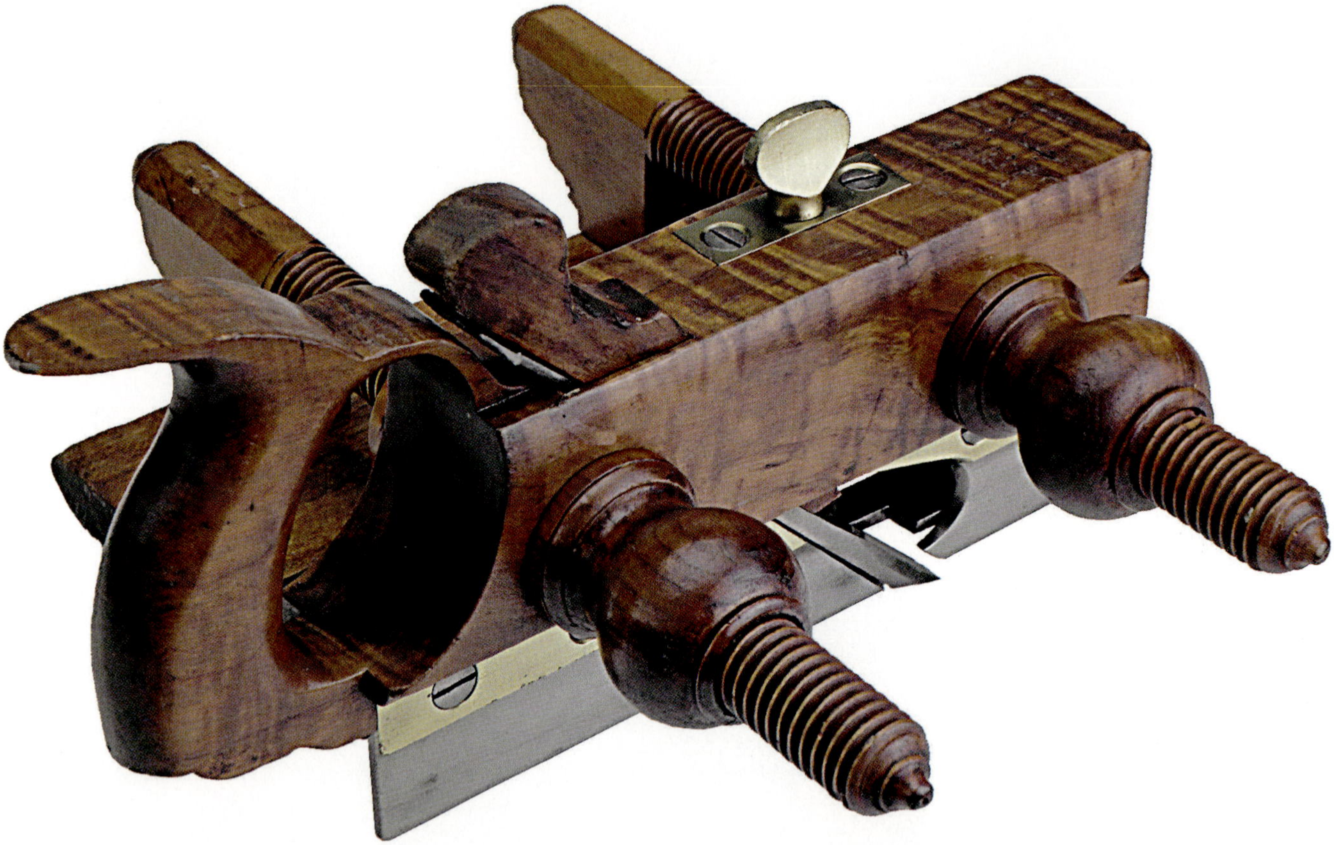

Edward Carter

Handled highly-figured beech screw-arm plow, with boxwood arms and flat nuts.

(From the collection of Don Rosebrook)

Edward Carter made planes under his own name in Troy, New York from 1854-1861 and 1865-1897.

At least one other plow with this strong figure is known. The large nuts are also made of beech, probably from the same piece as that used for the rest of the plane.

Benton, Evans & Co.

Unhandled beech screw-lock slide-arm plow with boxwood screws.

(From the collection of Mickey Holmes)

This Rochester New York planemaker operated from 1834-1838 and was a partnership between James Benton and Evan Evans. After 1838, these two men each went to work for a different plane maker.

L. Case

Unhandled beech screw-arm plow. B1 mark **

(From the collection of William Curtis)

Lorenzo Case made planes in Watertown, New York from 1850-1855. This plane is missing its fence. An L. Case plane is also known that has donut-shaped flat nuts as shown on p 167 for a Casey, Clark & Co. plane.

R. J. Collins

Unhandled boxwood screw-arm plow with ivory tips. ****

(From the collection of Ken Thomas >>>??)

Robert Johnson Collins made planes in Rochester, New York, beginning sometime between 1807 and 1809 and ending probably about 1827.

Robert Collins was among the first planemakers to incorporate ivory tips on screw-arm planes. Screw-arm planes were not being produced much before 1800 and those planes were typically made in Pennsylvania.

Martin Doscher

Handled boxwood screw-arm plow with depth stop lock.

(From the collection of Peter Hathaway)

Martin Doscher was a New York City hardware dealer from 1879-1894.

The nose and the base of the handle on this plane give it a blocky appearance.

A. L. Gleason

Handled rosewood screw-arm plow with boxwood fence, arms and nuts, and a depth stop lock. **

(From the collection of William Curtis)

Albert Gleason worked as a planemaker in Watertown, New York from 1859-1869. Pollak indicates that part of this time he worked alone, probably after 1863.

Gleason & Wood

Handled beech screw-arm plow with boxwood arms and nuts, and depth stop lock. **

(From the collection of William Curtis)

Pollak does not have definite information about the exact working period for this partnership, but the apparent period was from 1859 or 1860 until 1863. It was located in Watertown, New York.

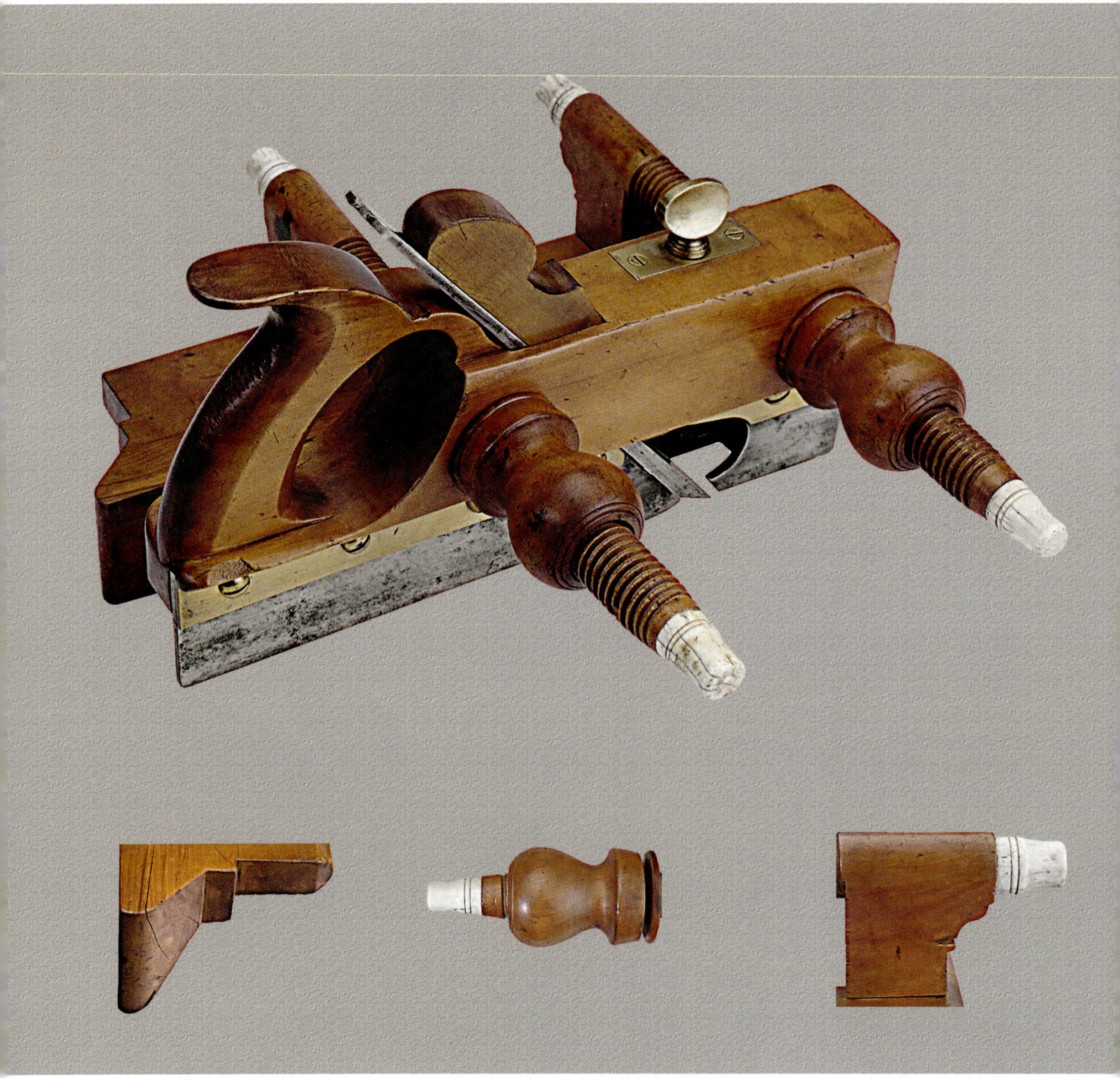

Gregory

Handled boxwood screw-arm plow with boxwood washers and ivory tips. **

(From the collection of Andrew Delans)

G. W. Gregory operated a hardware store in Binghamton, New York about 1850.

The ivory tips mark this plane as a Casey, Kitchell & Co. plane. The body is also from that maker, as the area at the base of the handle shows. The nuts and washers are still further indication of the Casey, Kitchell & Co. origin. The only question is the origin of the fence. It is double ended, but does not have the molded edge, typical of fences made by the prison planemaker. It is always possible that Gregory made his own fence, or that this is an after-market fence or, perhaps it was special order (but that seems unlikely since fancier fences should sell better.)

Randall & Bensen

Unhandled beech screw-arm plow with boxwood arms and nuts. Arms have acme threads. **

(From the collection of James Hill)

Samuel Randall and David Bensen made planes in Albany New York during 1827-1829.

Acme threads (recognizable by their flat top) were later used by Bensen & Crannell and still later by M. Crannell. These makers are among the few who utilized this thread type.

Scovill

Unhandled beech wedge-lock slide-arm plow with iron depth stop and brass collars on the arms. **

(From the collection of Don Rosebrook)

Lyman Scovill made planes in Johnstown, New York from approximately 1800 to approximately 1840.

Scovill

Unhandled beech wedge-lock slide-arm plow with beech depth stop and boxwood depth stop lock, unplated. **

(From the collection of Don Rosebrook)

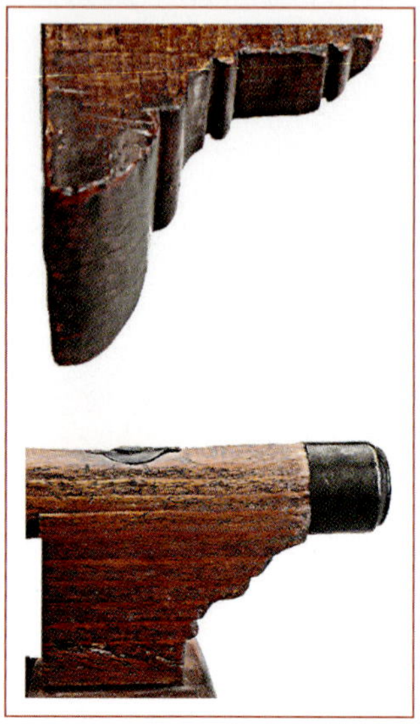

C. S. See

Unhandled beech wedge-lock slide-arm plow with round arms and boxwood wedges.

(From the collection of William Curtis)

Cornelius See was a plane maker in New York City from 1829-1846. He also is thought to have made planes in Watertown, New York although the dates are not known. Note the unusual round arms and the brass caps on ends of the arms. The use of brass caps is rare.

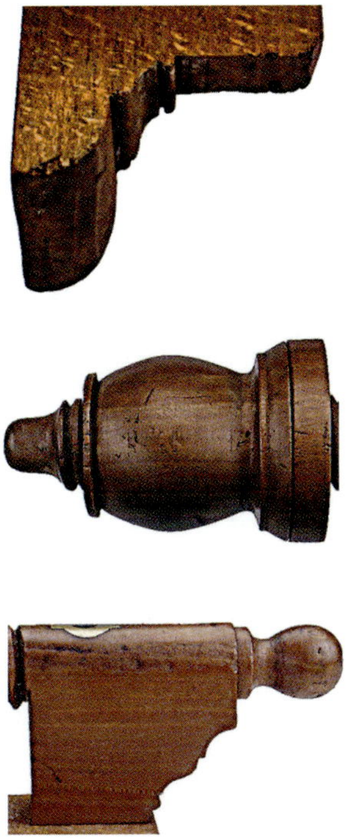

N. Spaulding

Handled beech screw-arm plow with added beech arm tips.
A mark *

(From the collection of Louis Read)

Nathaniel Spaulding made planes in McLean, New York from 1824-1850 and at the Treman & Bros. Ithaca, New York factory site from 1867-1870. The round finials, or tips, on the shoulder ends of the arms are extremely unusual.

Treman & Bros.

Unhandled boxwood screw-arm plow. **

(From the collection of Milt Bachelor)

Treman & Bros. was a hardware company in Ithaca, New York that operated under that name from 1849-1857.

De Valcourt

Unhandled mahogany screw-arm plow with boxwood fence, arm and nuts with rosewood washers and a depth stop lock. **

(From the collection of Frank Kosmerl)

Charles De Valcourt made planes from 1827-1836 in New York City at 107 Elm Street. According to Pollak, he subsequently became a grocer. Note the very fine threads on the arms. They look as though they were cut with a die for cutting metal threads.

L. Bundy

Dual handled maple screw-lock, slide-arm plow and match plane with mahogany handles and brass thumb screw for arm locks. Brass plaque on one side reads "L. BUNDY'S COMB'D MATCH PLANE AND PLOW PAT'D NOV. 15, 1870 MOORES FORKS N. Y." *****

(From the collection of Don Rosebrook)

Lewis Bundy, from Moores Forks, New York, was granted a patent on November15, 1870 for the combination match and plow plane. Only four examples of this plane are known and, according to Pollak, these may have been prototypes.

This is about as impractical as a plow plane might be. Although it will cut tongue and groove as advertised, the width of the tongue can be quite large but the width of the groove is limited by the width of the slot for the blade. The use of this as a plow plane is equally limited unless special grooving blades are made with narrow tops to fit in the slot.

Alford

Unhandled beech wedge-lock slide-arm plow with boxwood wedges, brass collars on the arms, and boxwood wedges in the ends of the arms. **

(From the collection of Mickey Holmes)

Consider Alford worked from 1812 to 1817 and is believed to have been one of the earliest New York City plane makers.

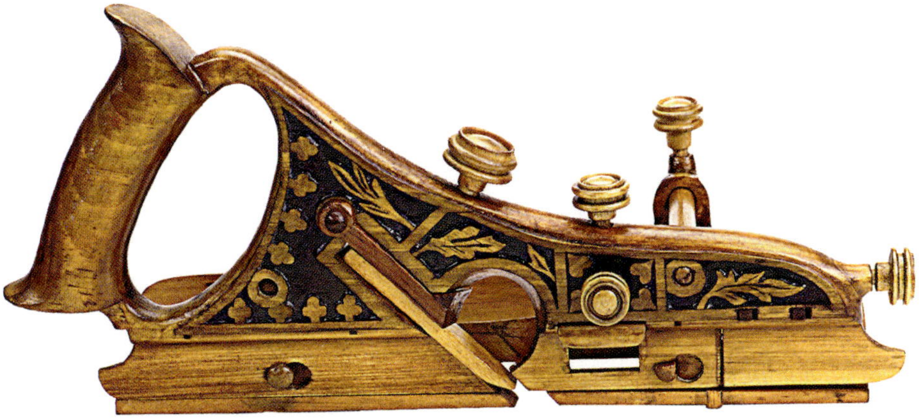

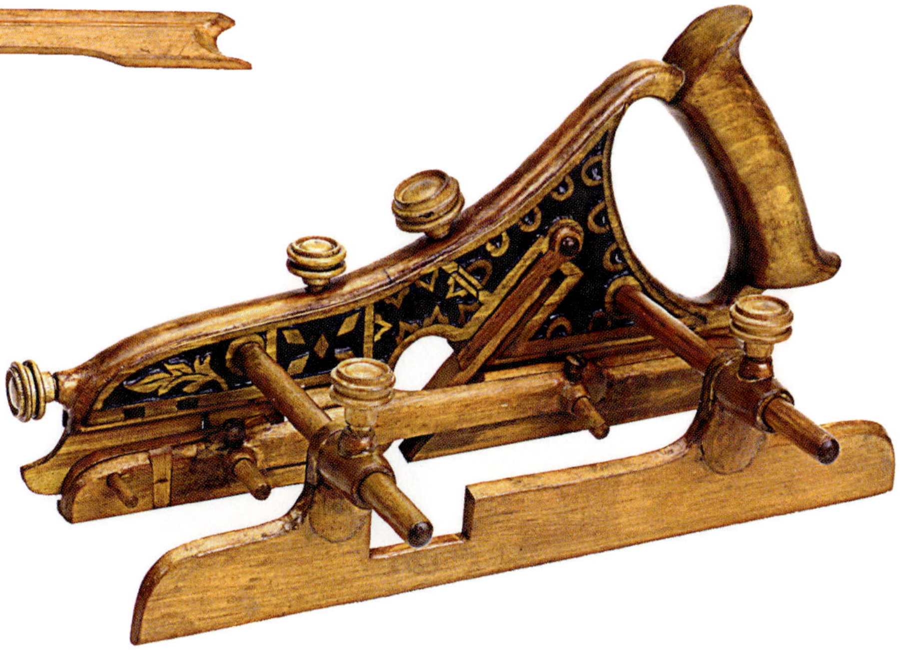

Siegley

Wooden model of a handled metallic plow plane. Patent model?

(From the collection of Mickey Holmes)

Jacob Siegley did not make wooden planes but did manufacture iron planes and held several patents for them. This wooden plane is possibly a model for one of his patents. Jacob Siegley lived in New York City from 1878 to 1883, and he was there when granted his first patent in 1879. That patent described a plane that closely resembles this model, although the manufactured plane has a very plain stock and fewer accessories. For a picture of this plane and more information see, Smith, Roger K., Patented Transitional and Metallic Planes in America 1827-1927, *North Village Publishing Co., Lancaster, Mass. 1981 p. 107.*

Andruss

Unhandled beech screw-lock self-regulating plow with iron wear strip on locking arm, brass collars on arms, double dovetailed boxing on the fence, and ebony wedges in the ends of the arms.
New mark like B with "warranted" added.

(From the collection of Mickey Holmes)

David Andruss and his brother George made planes and coachmaker's tools in Newark, New Jersey from 1821-1841.

This plane can be said to be made in the Mockridge & Francis style. It does, however, lack the lock nut on the outboard end of the center screw. To lock the fence on this plane there is a screw-lock on the forward arm and arm contains a wear strip. Otherwise, this plane is outfitted identically to some of the known Mockridge & Francis planes.

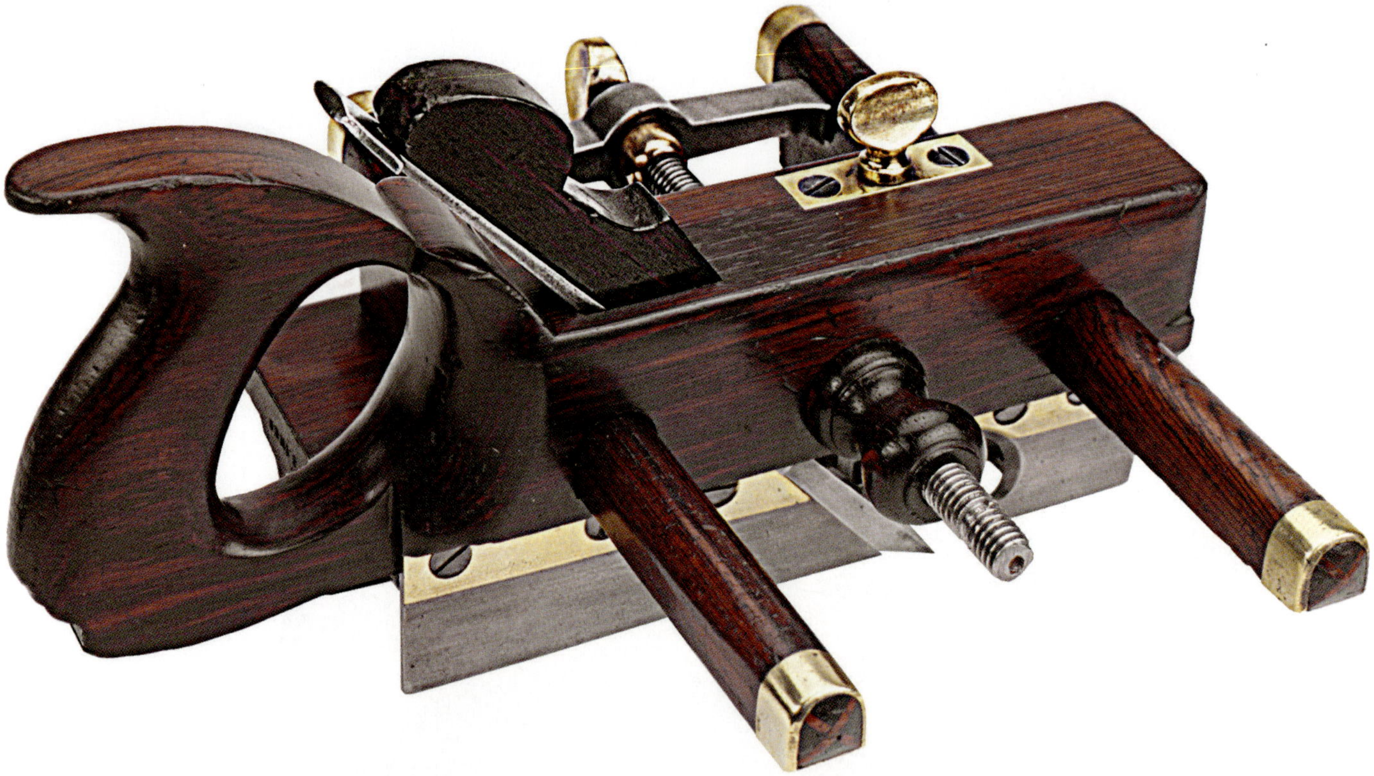

Mockridge & Francis

Handled rosewood self-regulating plow with brass collared arms, and boxwood wedges in the ends of the arms.

(From the collection of Mickey Holmes)

Abraham Mockridge and Elias Francis were partners in planemaking in Newark, New Jersey and in a hardware dealership from 1835-1868.

This is the more common arrangement of the Mockridge & Francis self-regulating plow. It has a threaded nut on the fence side of the body and a wooden nut on the outboard side of the body. The cast iron bridge-piece does not have any apparent support within the arms. This arrangement has been described by Herb Kean and he contrasted it to the Andruss plow. See Kean, Herb. "Three-Arm Plows." The Tool Shed 107 (1999)*: 4.*

Mockridge & Francis

Handled beech self-regulating plow with brass collared arms, boxwood wedges in the ends of the arms, and large brass lock nut on the center screw.

(From the collection of Don Rosebrook)

The large brass plate on the right hand side of the plane covers not only the threaded nut, but a variety of screw holes, showing that at one time the plane had different shaped cover plates. The solid brass "door knob" nut seems to be unique to this particular Mockridge and Francis plow. The cast iron bridge is supported like the bridge in the Andruss plane shown earlier, and this is contrary to the Kean article noted on the previous page.

MOCKRIDGE & FRANCIS,

Manufacturers of Planes,

AND Wholesale & Retail DEALERS IN

BUILDERS' HARDWARE

AND MECHANICS' TOOLS,

No. 235 (Old No.211,) WASHINGTON STREET, NEWARK, N. J.

S. C. Cook

Unhandled boxwood screw-arm plow with depth stop lock, brass collars on the ends of the arms, and rosewood wedges in the ends of the arms.

(From the collection of Don Rosebrook)

Samuel Cook made planes in New Brunswick, New Jersey from 1825-1845. This plane has wooden diamond-shaped plugs on the top of the arms. The authors have not seen this on other planes.

Maryland & Pennsylvania Planemakers

3

Maryland was home to several planemaking enterprises, nearly all of which were located in Baltimore. Their influence upon one another is obvious from the few Maryland planes shown here.

Pennsylvania hosted many planemakers, the most famous of which were located in Philadelphia and Lancaster. Israel White of Philadelphia and E. W. Carpenter of Lancaster were each holder of a patent for plow planes, granted in the 1830s. Their planes, in particular the patented planes, are sought after by collectors from all over the United States, but with greatest fervor by collectors in eastern Pennsylvania. Their plow planes had a special charm and beauty about them as is demonstrated by the photographs in this section.

The center screw fence adjustment, invented by Israel White, was often imitated by other planemakers, including one from Cincinnati recently discovered and not shown here. In particular, there are many craftsman-made planes that utilize the principles of Israel White's patent.

Two Pennsylvania planemakers who were immigrants from England, John Butler and Thomas Napier, worked in Philadelphia, beginning in 1891 and 1874, respectively. They are among the earliest planemakers outside of New England and are thought to have been making screw-arm planes as early as 1800-1810. This would make them the first makers of American screw-arm planes, insofar as we know.

The use of ivory and silver trim was much more limited in Pennsylvania and Maryland than in other sections of the country.

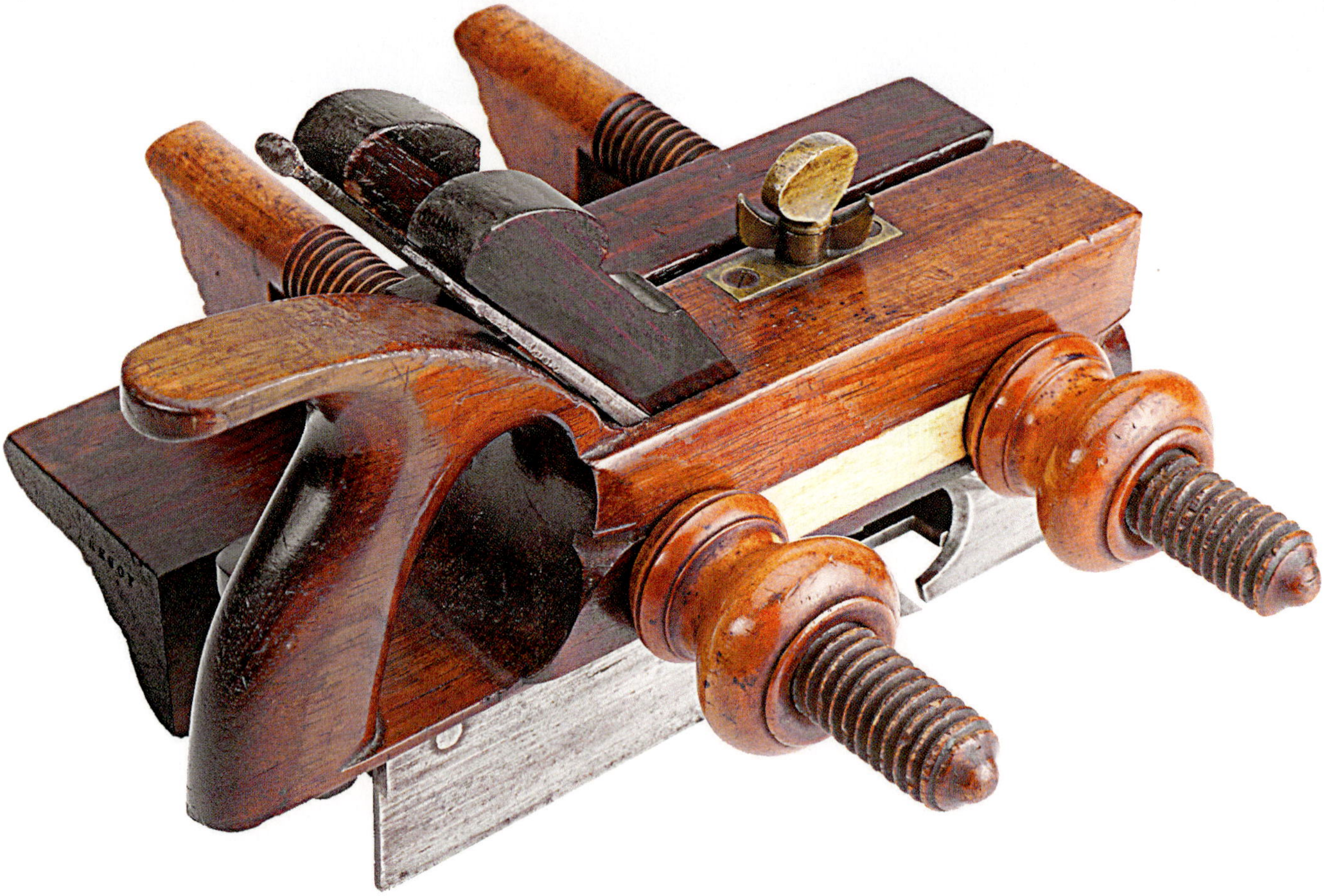

H. L. Kendall

Handled rosewood screw-arm plow with boxwood arms and nuts. Plane has extra rosewood body on same set of arms, ivory wear strips on the face of fence, ivory escutcheons for the screws on the bottom of the fence, and a decorative ivory plate on the handled body piece. Locking wing nut on the depth adjustment screw. **

(From the collection of James Hill)

Henry Lee Kendall was a hardware dealer and planemaker operating in Baltimore from 1849-1860. There appears to be some connection between this Kendall and Robert MacCubbin, although Pollak states that Thomas Kendall and MacCubbin were in partnership from 1837-1839.

Jim Hill, in his article in The Chronicle *entitled "The Piggyback Plow," discusses the existence of at least three planes of this style. (Hill, Alexander James III. "The Piggy Back Plow."* The Chronicle *35 (1982): 80.) One is the plane shown here, while the other two are planes signed P. Chapin, another Baltimore planemaker. Among the characteristics Mr. Hill emphasizes are the closeness of the arms — approximately 1/2" closer than normal, and the length of the arms — approximately 3/4" longer than normal. The use of a lock screw immediately under the thumb screw for the depth stop is very unusual and also seems to be primarily characteristic of Baltimore planemakers. Other methods of locking the depth stop can be found on New England planes.*

R. W. MacCubbin

Unhandled beech screw-arm bridge plow with boxwood arms and nuts. *

(From the collection of Emil and Martyl Pollak)

Robert MacCubbin was a Baltimore, Maryland planemaker who was listed as a partner in Kendall & MacCubbin and as a planemaker in 1840, 1845-1858 and 1865-1877.

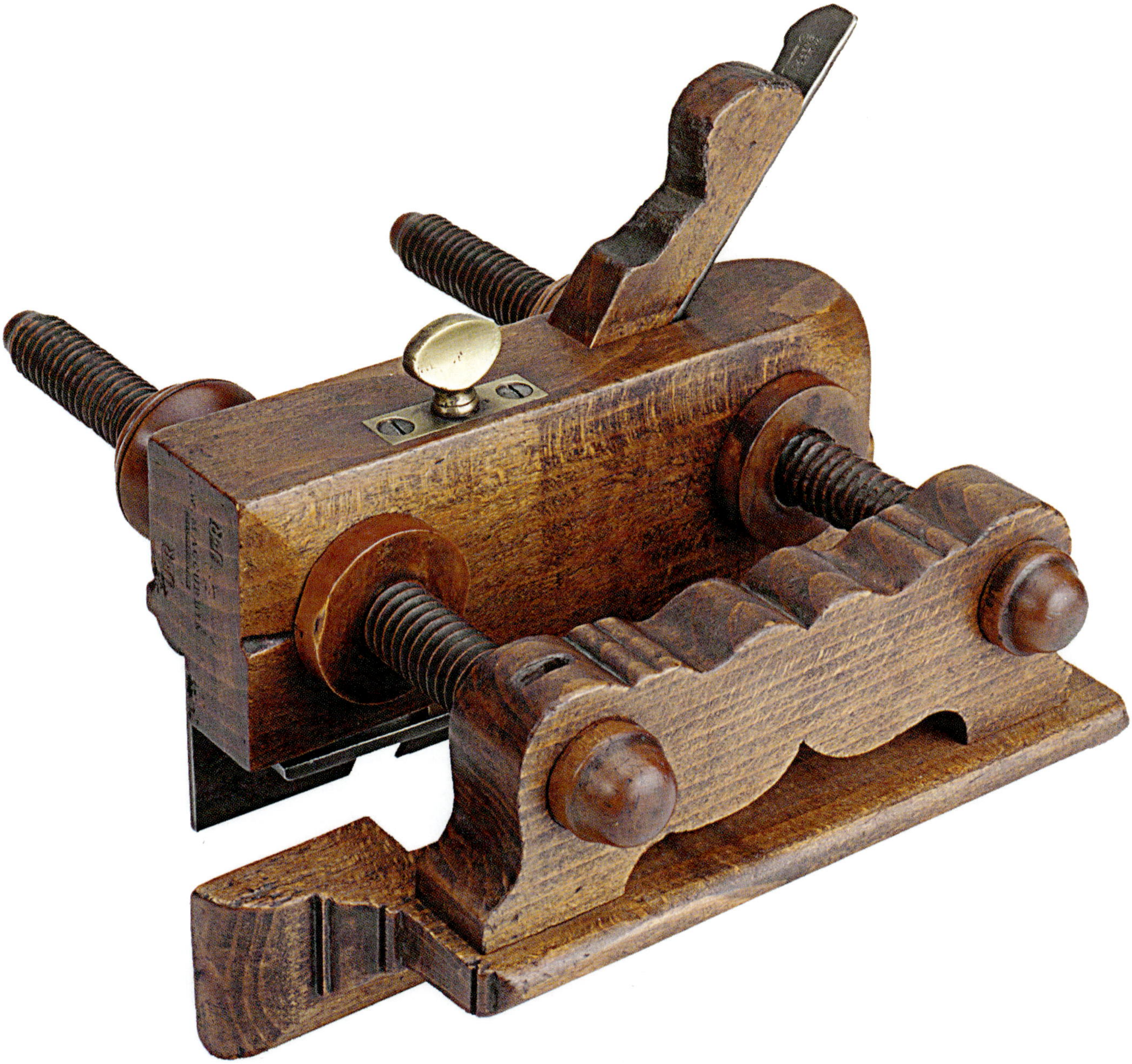

Screw-arm plows with a bridge on the fence are quite rare. The authors believe that the designs on the bridge will be characteristic of the maker. Compare the bridge on this plane with that of the unsigned plane on the next page.

R. W. MacCubbin

Unsigned, handled boxwood screw-arm bridge plow with ivory inlay in bridge and ebonized holly inlay on body of plane. Locking wing nut on the depth stop adjustment screw.

(From the collection of Don Rosebrook)

The dark blotches in the boxwood are original to the wood and are not stains.

The resemblance of the bridge on this plane to that on the MacCubbin plane is remarkable. The nuts and fence molding seem to more closely resemble those on the MacCubbin plane than those on the Kendall plane. The fence has a brass facing strip and a brass plate under the bridge. These both seem original to the plane. The brass facing on the wedge may not be original to the plane. The extensive decoration suggests that this was a special order plane, perhaps intended for a presentation piece. This plane utilized hand-made screws which should date it to the early part of MacCubbin's working period. It is a generally accepted notion, but not an established fact, that the use of hand-made screws stopped abruptly in 1846 after the invention of a machine to point screws.

When this plow was found, the arms were secured to the bridge with brass machine screws with square nuts on top. During restoration these screws were replaced and the nuts were replaced by threaded inserts. Restoration by Robert Baker.

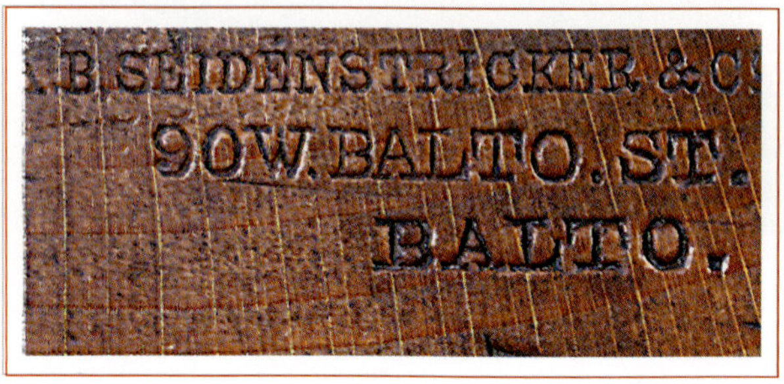

A. B. Seidenstricker & Co.

Handled beech screw-arm bridge plow with boxwood arms and nuts and locking wingnut under the depth stop adjustment screw.   C mark **

(From the collection of James Hill)

A. B. Seidenstricker & Co. made planes and sold tools in Baltimore, Maryland from 1856 to 1866. A. B. Seidenstricker was probably active both before and after those dates as part of a partnership or another company.

The bridge on this plow is quite different from that on the MacCubbin plow, but the ends of the arms at the bridge and the molding design under the handle speak strongly of a kinship with the unsigned plow on the previous pages. The lateness of the Seidenstricker operation suggests that he was not the maker of this plow. Perhaps it is a later MacCubbin plow that Seidenstricker had marked as a hardware dealer rather than as a maker. Finally, we do not have a P. Chapin plow against which to compare these plows. It may well be that P. Chapin (probably Seidenstricker's mentor) made both the unsigned plow and this plow or that P. Chapin made the unsigned plow and passed on the techniques and patterns to Seidenstricker.

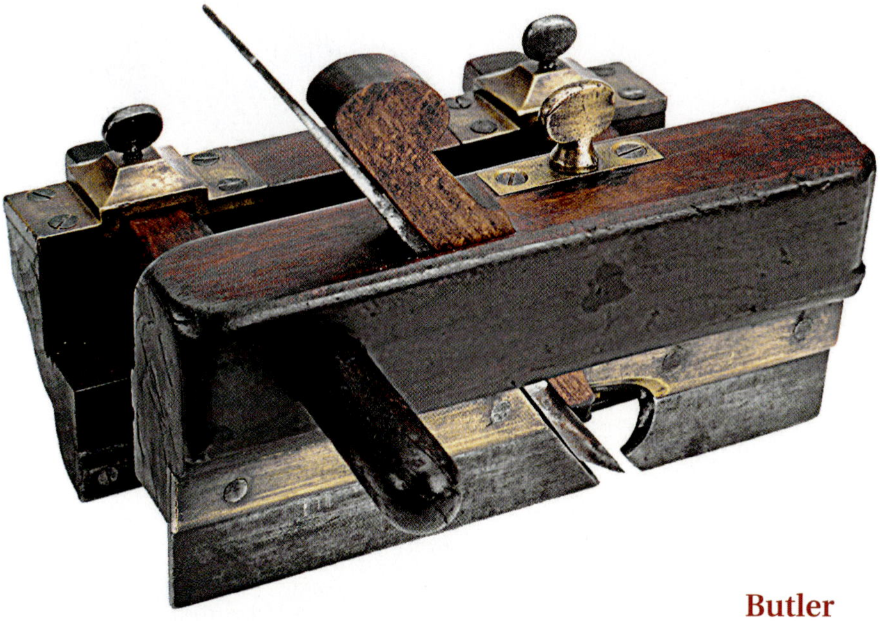

Butler

Unhandled beech screw-lock, slide-arm plow. Arms fixed to body not fence. B mark *

(From the collection of Mickey Holmes)

John Butler is known to have worked in Philadelphia from as early as 1791 to as late as 1830. This might be called a quasi-continental style plow plane because it has the arms fixed to the body. It is an extremely rare style for an American plane.

E. W. Carpenter

Unhandled beech screw-arm plane with boxwood arms and nuts, and oversize wedge. Skate screws go through the body and are cut flush on the right side, plus a large rivet through skate in center.

(From the collection of Mickey Holmes)

This is a very small plow plane. The body is only 6½" long. Normal body length is 8" to 9".

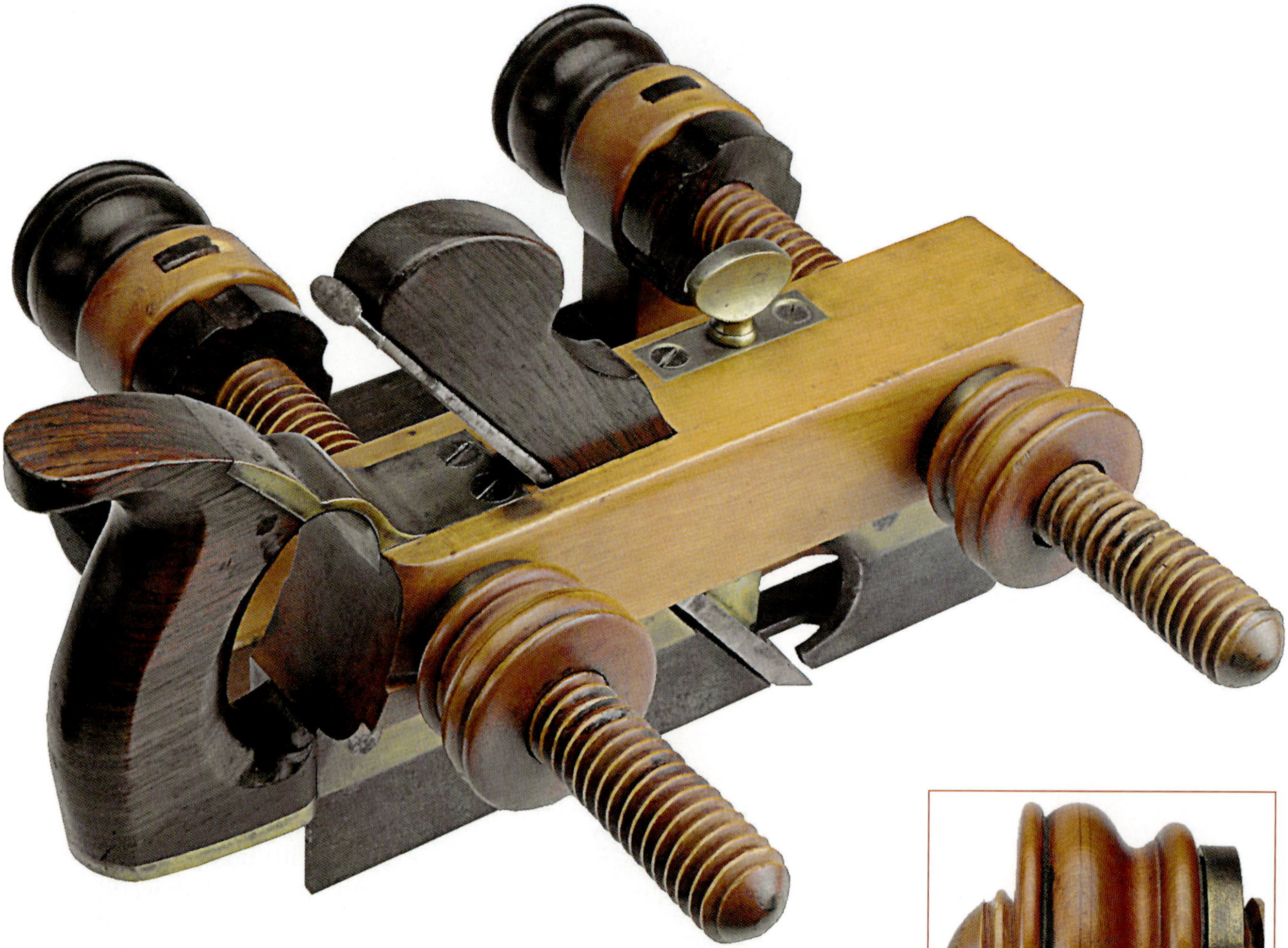

E. W. Carpenter

Handled boxwood screw-arm plane incorporating the "improved arms" patent of February 6, 1838. Rosewood fence with massive boxwood boxing, oversize rosewood wedge, rosewood handle, boxwood arm stanchions and rosewood knobs on arms. Skate held with large rivets. New mark like D2****.

(From the collection of Don Rosebrook)

Emanuel Carpenter opened his plane making business in 1826 in Lancaster, Pennsylvania and continued there until his death in 1856. His wife Sarah ran the business after his death until 1863.

Carpenter held three patents for planes, the second of which was for the "improved arms" plane shown here. Bulky wedges are a distinctive characteristic of his planes.

D. Colton

Handled beech screw-arm plow with (fruitwood or satinwood) arms and nuts, and massive rosewood boxing. C mark *

(From the collection of Mickey Holmes)

David Colton was a Philadelphia planemaker active from 1837 to 1880. Part of that time he was in partnership with his brother John, and part of the time (1846-1852) he was in partnership with Benjamin Sheneman.

Colton worked at a number of locations within Philadelphia, as reflected in the name stamps that he used.

Goldsmith

Handled ebony, screw-arm plow with boxwood arms and large rosewood nuts, flat nuts of rosewood and bone, and silver mounting.   A mark *

(From the collection of Roy Schaeffer)

William Goldsmith was a planemaker in Philadelphia. He worked from 1837-1868. He was located at the corner of New Market and Green Street, as his stamp indicates.

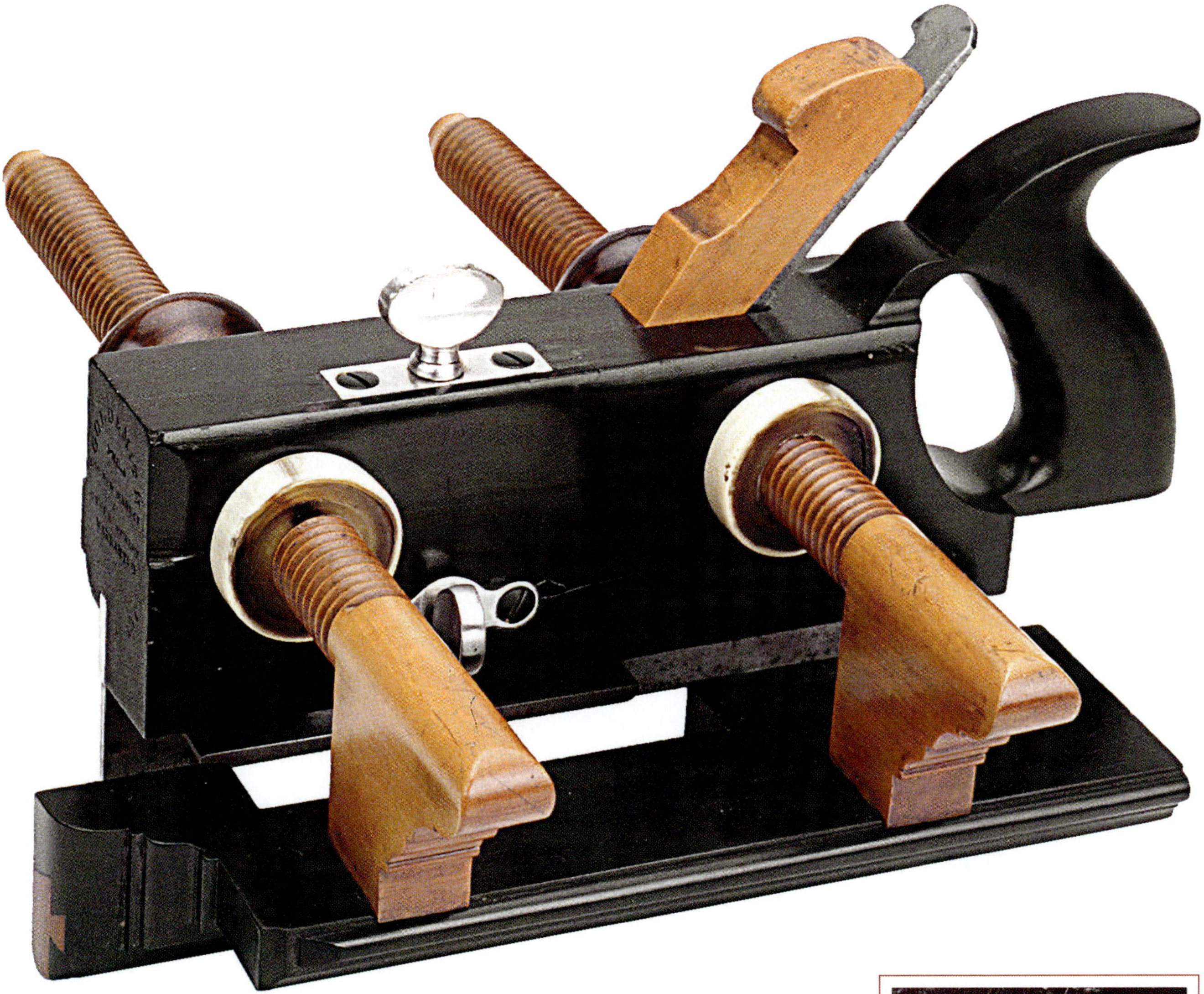

The nuts on this plane are most unusual in construction and in the use of bone in such a high end plane.

B. Sheneman

Handled beech screw-arm plow with rosewood nuts and wedge, and rosewood boxing on the fence.
(From the collection of Mickey Holmes)

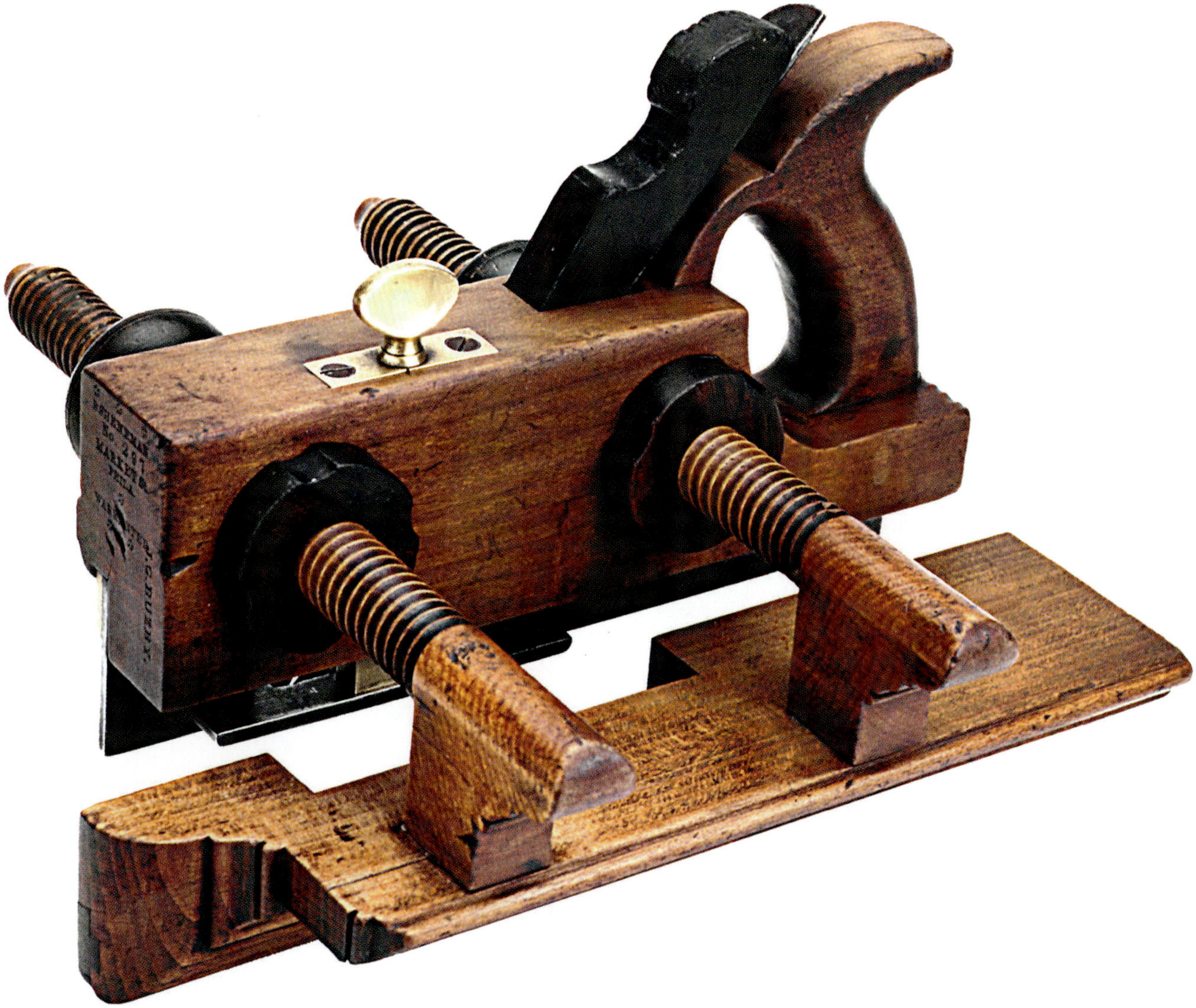

Benjamin Sheneman was a planemaker, working at various locations in Philadelphia, Pennsylvania from 1846-1867. During that time, he worked alone from 1852-1856 and from 1860-1867.

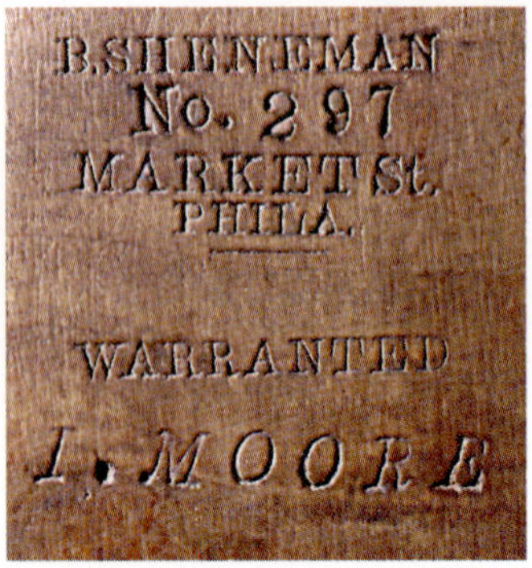

B. Sheneman

Handled fruitwood, screw-arm plow with rosewood wedge, mahogany nuts, and boxed fence.
(From the collection of Mickey Holmes)

Note that the shape of the flat nuts is very different from that of the previous B. Sheneman plane.

Mr. Haines, Mr. Pilgrim, and Mr. Moore took no chances in having their plane mistaken for someone else's.

T. J. Sheneman

Handled beech screw-lock slide-arm plow, with brass escutcheons for arm-lock thumb screws. New mark.

(From the collection of Mickey Holmes)

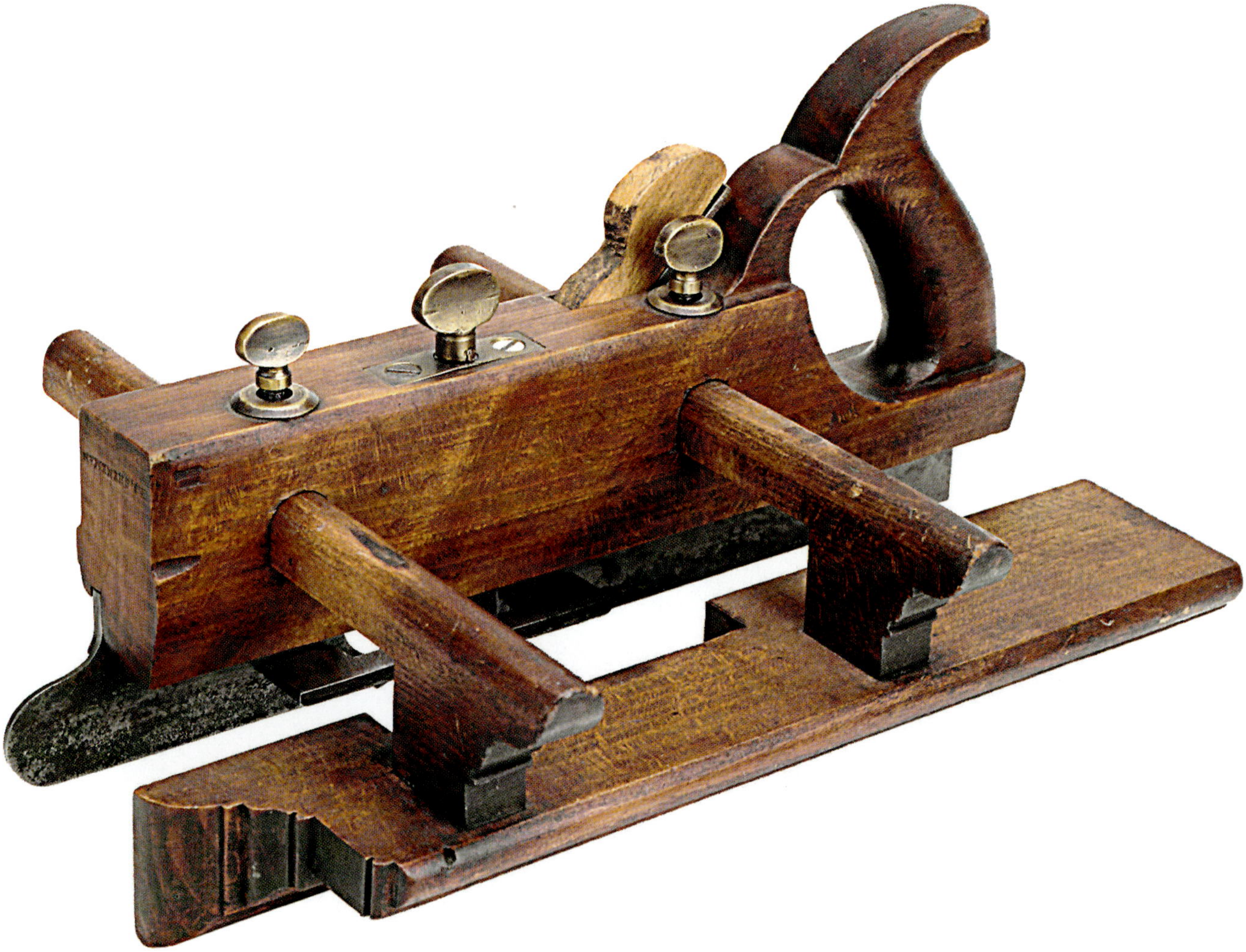

This imprint is not recorded by Pollak and T. J. Sheneman is not listed as an individual planemaker by Pollak. However, he is known to have worked with his brother Benjamin from 1856-1860.

G. White

Unhandled beech screw-arm plow with captive nuts and boxed fence. New mark.

(From the collection of Mickey Holmes)

George White was the brother of Israel White. George apparently operated under his own name from 1820-1824. He died in 1824 at the age of 33 and his widow, Catherine, continued the business until her death in 1825. It is believed that his father, Jacob, carried on the business using George's mark until ca. 1831 when Israel took over.

Israel White

Handled beech screw-arm plow with boxwood arms and nuts. Bench hand mark A M for Adam Miller.

(From the collection of Mickey Holmes)

Israel White was the son of Jacob White. Pollak believes that he was a journeyman in his father's shop from 1825-1831. Pollak speculates that he did not imprint planes with his own name until his father's death in 1833. He made planes under his own name in Philadelphia from 1833 until his death in 1839. His widow, Charlotte, continued running the shop from 1839-1845. Henry G. White, Israel's nephew, took over the shop in 1845. He was the orphaned son of George and Catherine White.

Israel White

Unhandled beech, patented self-regulating plow with round ebony arms and wedge, ebony wear strip on fence, and boxwood center screw and nut. New mark ****

(From the collection of Mickey Holmes)

This plane was patented on January 9, 1834, which was very early in Israel White's career as an independent planemaker.

Israel White

Unhandled beech screw-arm plow with boxwood arms and nuts, and mahogany boxing on fence.
F mark ***

(From the collection of James Hill)

Note the strong resemblance between the nuts on this Israel White plane and those on his brother George's plane.

J. B. Mack

Unhandled beech screw-lock slide-arm plow, with iron thumb screws through iron plates for arm locks. **

(From the collection of Mickey Holmes)

John B Mack was a plane-maker in Brooklyn Township (Susquehanna County) Pennsylvania. He is believed to have worked ca. 1836.

4 Ohio & Kentucky Planemakers

Two of the largest planemakers in America were located in Ohio: the Ohio Tool Company in Columbus and the Sandusky Tool Company in Sandusky. Both companies operated during the last half of the 19th century and the first quarter of the 20th century. The product lines of the two companies were similar, as was the style of their plow planes. The ebony center-wheel plow planes of the two companies (see discussion of center wheel plow planes opposite) are among the most sought after of American planes by collectors and they are certainly among the rarest, with a total of three such planes known.

Kentucky had few planemakers and their operations were not large. This book shows a great example from each of two Louisville planemakers. These planes are believed to have been presentation planes and each demonstrates the considerable skill of its maker.

Ohio Planemakers and the Origin of the Center-Wheel Plow Plane[1]

The beauty and complexity of "self-regulating," or center-wheel, plows made by the Ohio Tool Company and the Sandusky Tool Company has always been fascinating. There is no known center-wheel plow plane marked with a patent date or related information. However, John Moody, in his *American Cabinetmaker's Plow Plane* (1981) published the drawings of H. Van Buskerk's patent self-regulating plow plane (Nov. 30, 1869). The design described in that patent is almost identical to both the Ohio and Sandusky designs. He then raised the question as to which company first offered this plane.

Based on evidence presented here, it now seems as though the Ohio Tool Company was not only the first to produce such a plane, but was the first by many years. The evidence suggests that the Ohio Tool Co. introduced and marketed the center-wheel plow in the 1850s. Tom Witte feels that the beech center-wheel plow shown in this chapter was among the very first of such planes. The beech plane is unmarked but the ivory tip design (center-balled), the styles and size of the body and fence, and the thumb screws are consistent with signed planes made by the Ohio Tool Co. There is also the use of the same odd thread count on the hardware as found on Ohio planes, and when this plane was found, it had an Ohio iron. There are several features on this plane not found on later pieces, probably because they were too costly to produce. These features include a center-wheel and threaded shaft turned from a single piece of wood, a lock nut placed within the confines of the handle, and the use of inexpensive and unimpressive wood to produce a plane of great complexity. (Future models were only to be made in boxwood and rosewood [the ebony plane shown in this section was found since Mr. Witte set down his thoughts.]).

At least one more plane of this style has been found and serves to reinforce the idea above. The second plane was made of boxwood, in what can be called the normal design. However, the center-wheel and threaded shaft are still in the one piece design used in the beech plane, and the bridge is still ornate, as are the mushroom-shaped thumb screws and the steel depth stop adjusting screw.

It is well known now that Ohio Tool Company made planes for resale by other companies under their own brands carrying only the Ohio Tool Co. model number. The boxwood plow referred to above is marked with the model No. 110, but carries the brand "Fall & Cunningham Nashville, Tenn." According to Pollak, this hardware store operated from 1853-1861, dating the manufacture of the Ohio center-wheels to somewhere in that period. This predates the establishment of the Sandusky Tool Co. and Buskerk's 1869 patent.

Could it be that Buskerk took the Ohio Tool Company idea and patented it? It would not be the first time that a patent was obtained for something that was not new. Could it be that Sandusky encouraged Buskerk to patent the plane so that they would have less of a problem copying the design? Ultimately, the only difference between the planes of the two companies is the relatively minor one of a brass center-wheel or a wooden center-wheel.

[1] This discussion is based on information provided by Tom Witte.

Ohio Tool Co.

Handled beech self-regulating center wheel plow with boxwood bridge, bushings, center wheel and screw, boxed fence, 6 ivory tips and depth stop lock.

(From the collection of Durward (Jim) Gehring <<< Thomas Witte)

There are several traits that mark this as an early Ohio Tool Company plane: the interesting little balls in the center of the ivory tips, the placement of the lock screw for the rear arm, and the ornate bridge. The interesting center wheel is noteworthy, being an integral part of the threaded center arm. See the discussion on center-wheel plow planes in the introduction to this chapter.

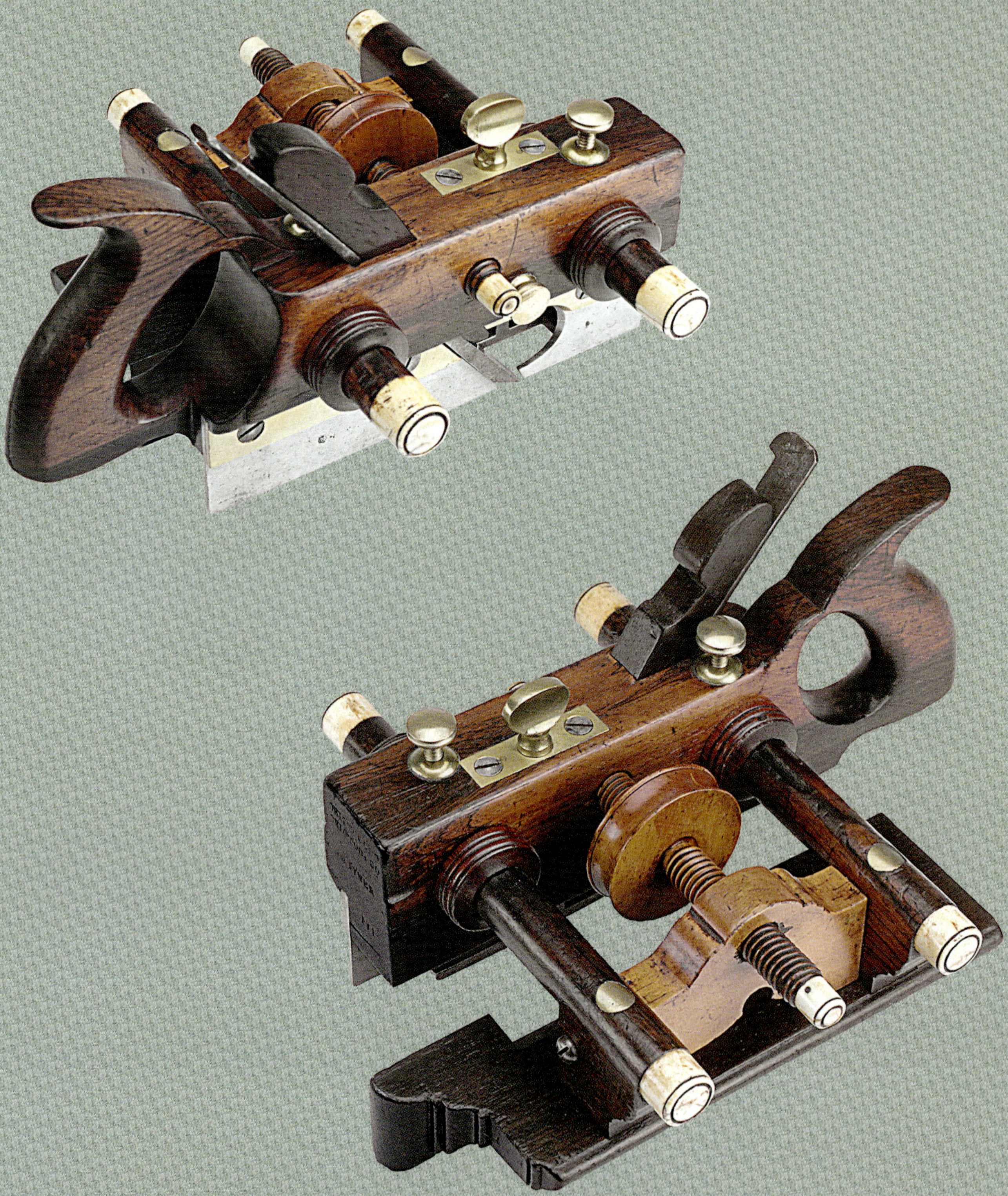

Ohio Tool Co.

Handled rosewood self-regulating center-wheel plow with boxwood center wheel and screw, six ivory tips, and depth stop lock. Model 111

(From the collection of Don Rosebrook)

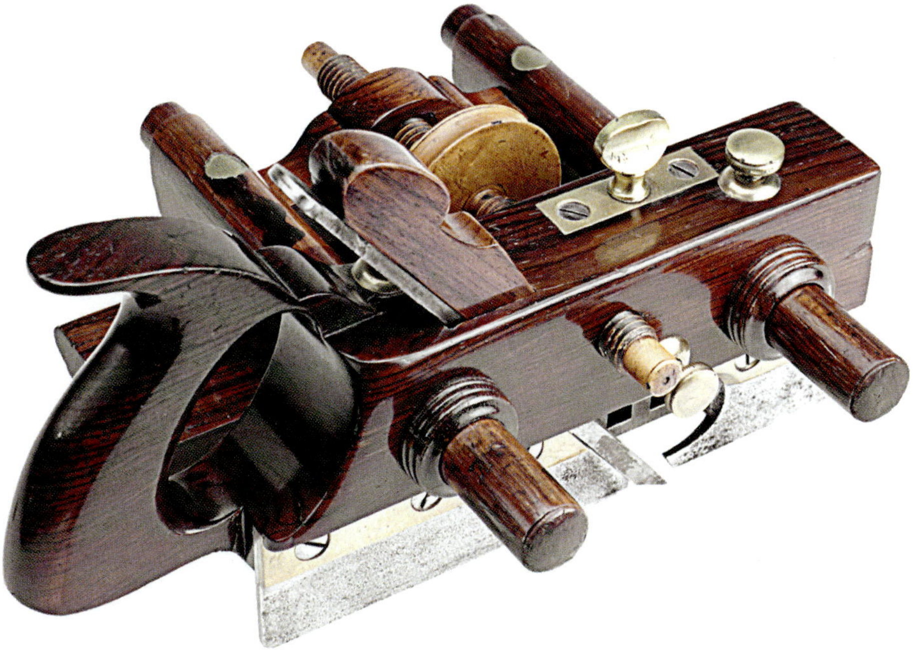

Ohio Tool Co.

Handled rosewood self-regulating center-wheel plow with boxwood center wheel and screw, and depth stop lock. Model 110

(From the collection of Durward (Jim) Gehring <<< Don Rosebrook)

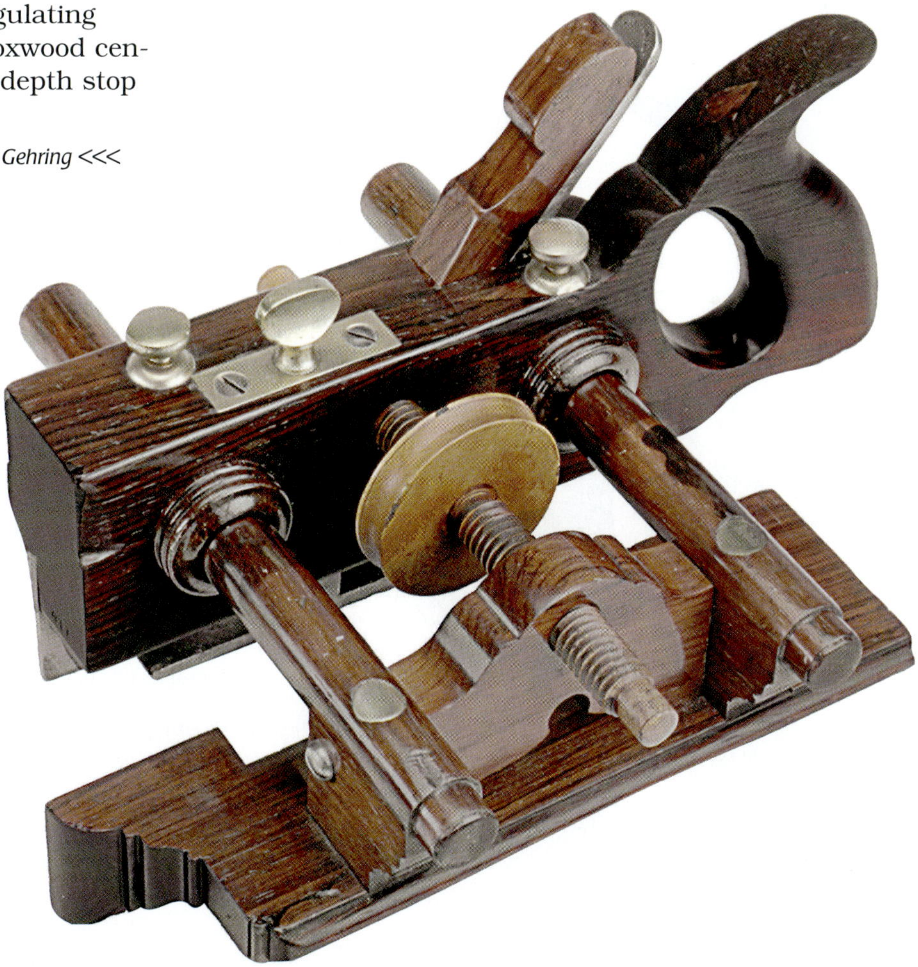

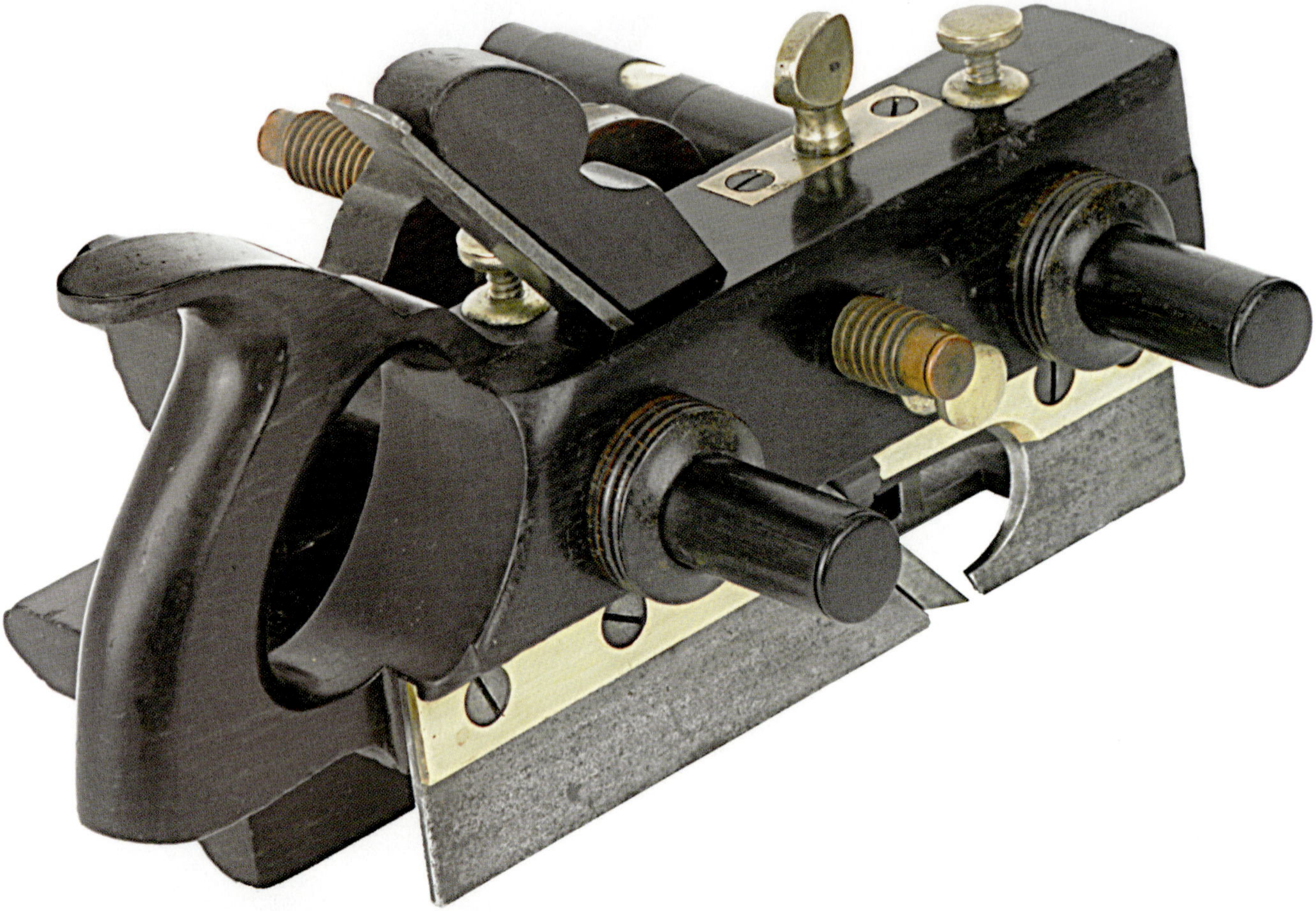

Ohio Tool Co.

Handled ebony self-regulating center-wheel plow with rosewood bridge, bushings and center wheel, boxwood screw, and depth stop lock. Model 112.

(From the collection of Don Rosebrook)
Photographs this page and next by Larry Weaver.

This ebony-center wheel plow is the only one known. The authors do not know of any Ohio Tool Co. ebony center-wheel plow with ivory trim. However, Sandusky ebony center wheel plows with ivory trim are known.

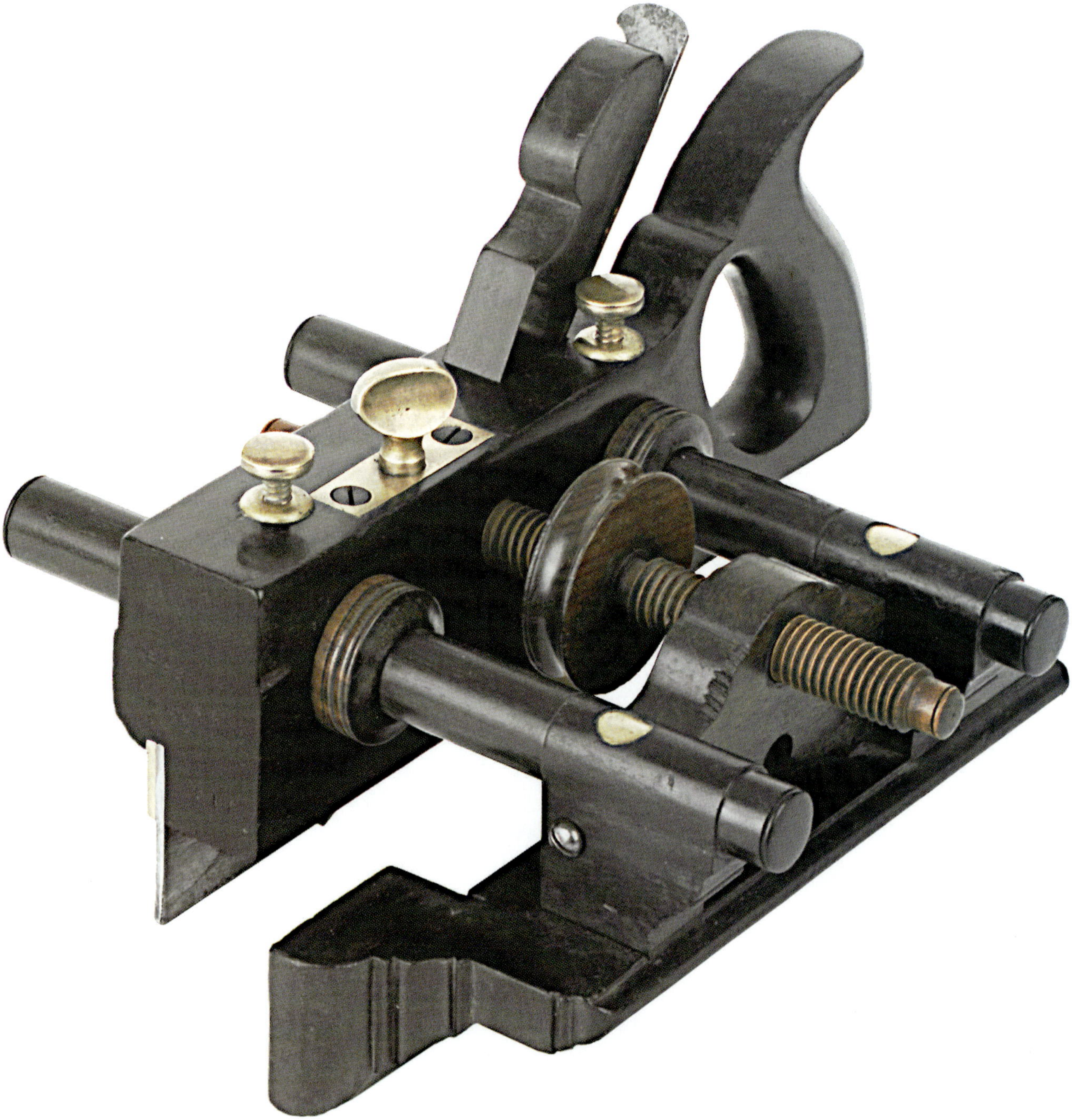

This plane was found in Pennsylvania in 2001. It was badly damaged, and was restored by Robert Baker. The use of rosewood in combination with ebony is an Ohio Tool Company characteristic, and while it may be functionally fine, it does not show as well as boxwood combinations.

Ohio Tool Co.

Handled ebony screw-arm plow with rosewood arms and nuts, ivory tips, and depth stop lock. Model 109

(From the collection of Don Rosebrook)

Ohio Tool Company was incorporated in Columbus, Ohio in 1851, although it may have operated as early as 1823. It used prison labor beginning in 1852 and later from time to time.

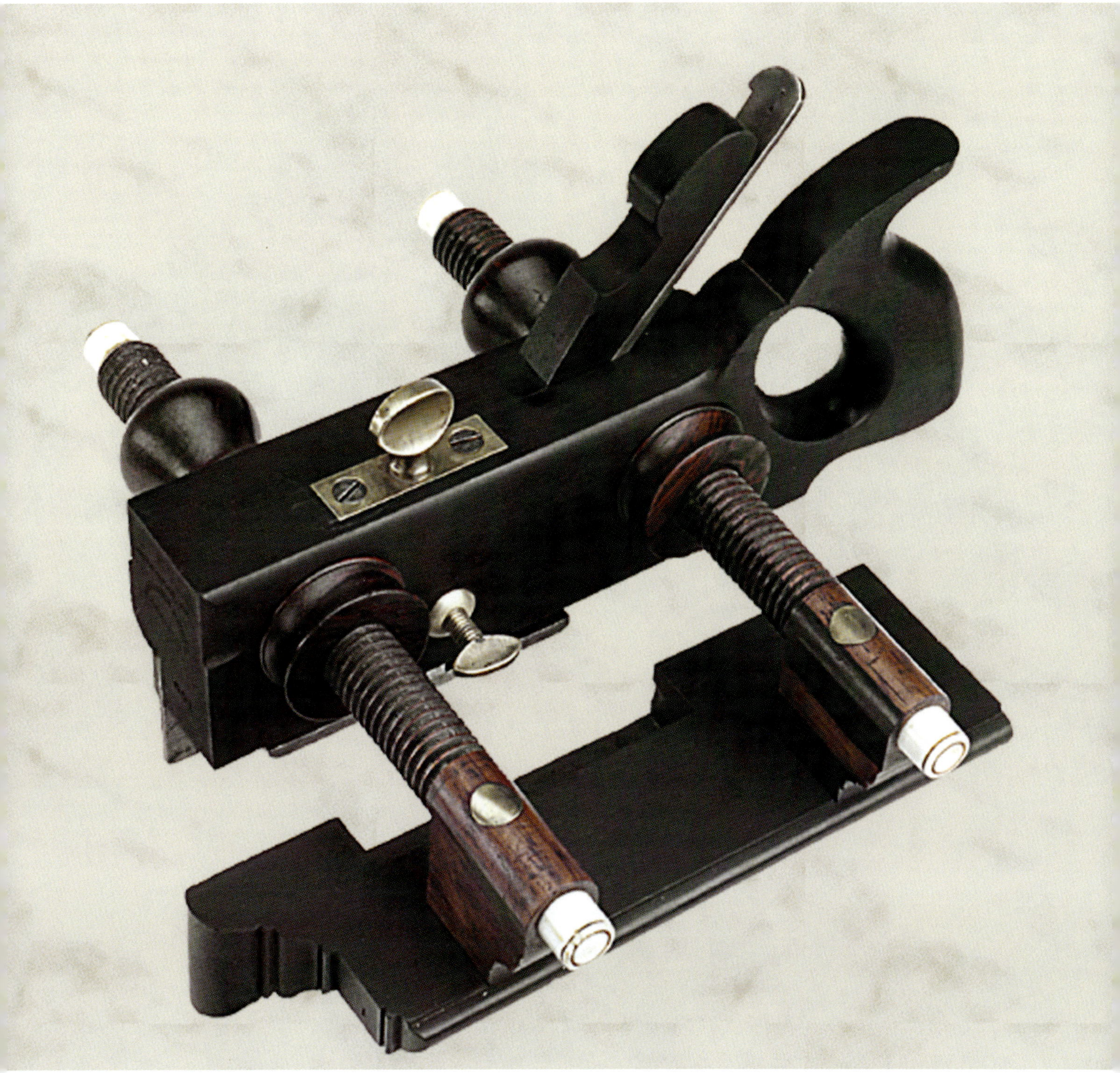

The Ohio Tool Company was a very large company. It was reported to be employing 200 hands in 1851 with carpenter's planes as the primary product. In 1893, the company merged with Auburn Tool Company and the combined operation retained the Ohio Tool Company name. The plane factory was destroyed by flood in 1913. Plane manufacturing was renewed in 1914 at a factory in Charlestown, West Virginia and continued through 1920.

Ohio Tool Co.

Unsigned, handled rosewood screw-arm plow with ivory tips and 6 pieces of ivory inlay. Skate is held on by round headed saw studs.

(From the collection of Thomas Witte >>> ??)

This fine rosewood plow with the ivory inlay is a great example of the plane makers art, but it was not always cherished. The inset pictures above and on the next two pages show what this plane looked like when it was found.

There are no identifying marks on the plane, but its features definitely mark it as an Ohio Tool Company product. It does have one very odd feature — the screws attaching the skate to the body are actually saw type studs.

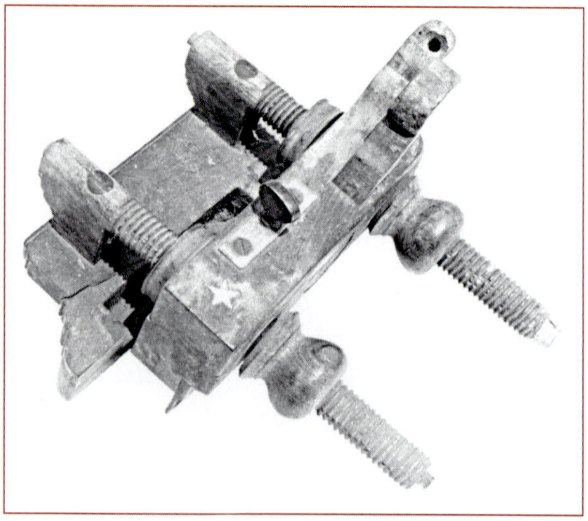

When the plane was found, the major parts were essentially intact and enough of the inlay was left to enable a proper restoration. This work was done by Robert Baker who is profiled later in this book.

E. F. Seybold

Unhandled beech screw-arm plow with rosewood nuts. B mark **

(From the collection of Mickey Holmes)

Emanuel Seybold is said to have made planes in Cincinnati, Ohio from 1836-1844. Other information presented by Pollak either states or implies that Seybold continued to make planes until his death in 1853.

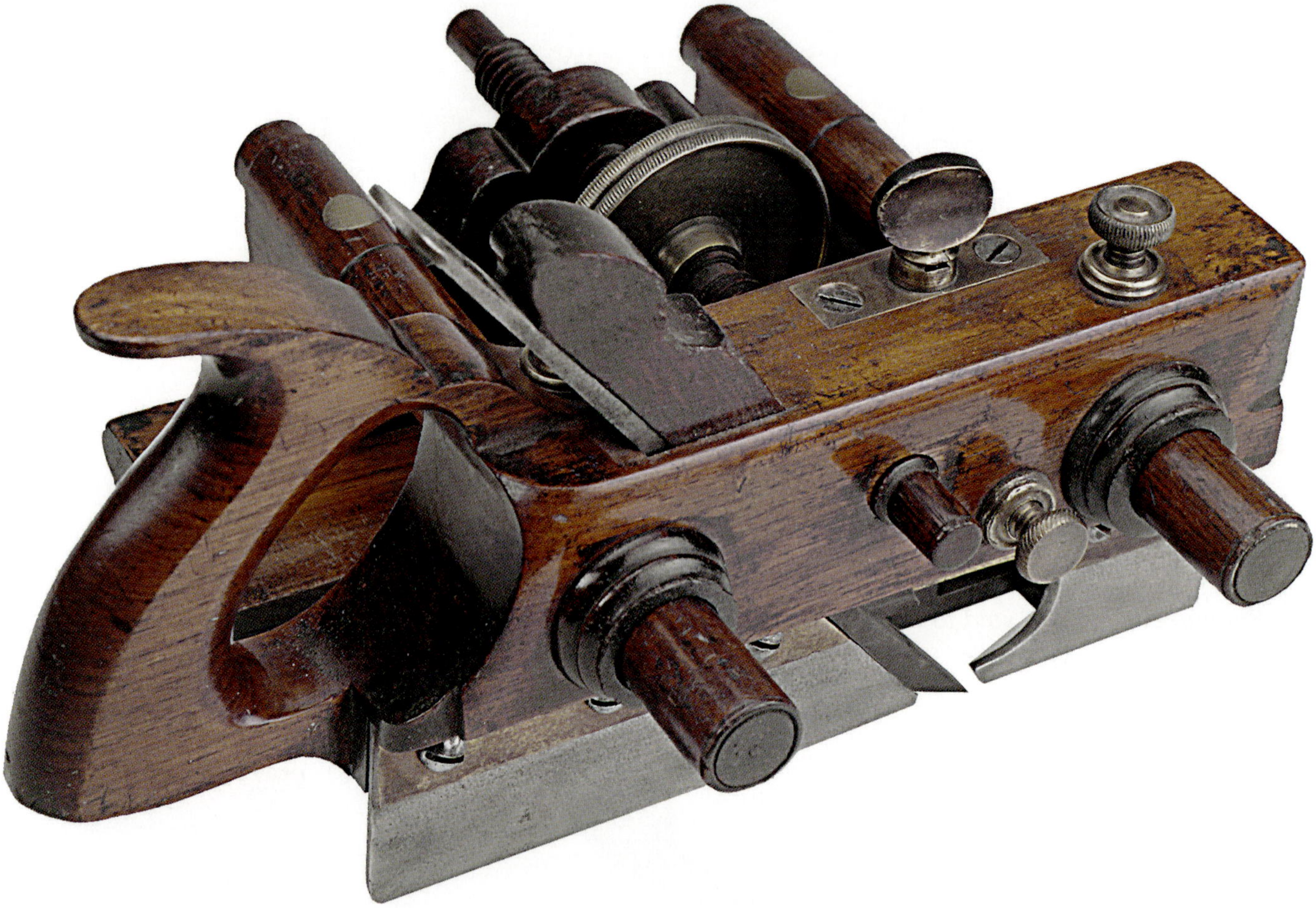

Sandusky Tool Co.

Handled rosewood self-regulating center wheel plow with brass center wheel and depth stop lock. Model #140

(From the collection of Frank Kosmerl)

The design and operating principles of this plane are identical to those of the Ohio center wheel plane. The two makers produced planes of a sufficiently similar style, that if someone replaced the wooden center wheel from an Ohio plane with the brass center wheel from a Sandusky plane, or vice versa, only an expert could detect the switch. The differences are subtle but significant and the reader may wish to compare the two rosewood center wheel planes presented in this chapter for their own edification. (Start with the thumb screws and grommets.)

Sandusky Tool Co.

Handled ebony screw-arm plow with ivory tips, fence screws going into t-nuts, and depth stop lock. Model 137

(From the collection of Don Rosebrook)

The Sandusky Tool Co. was founded in 1869 in Sandusky, Ohio. It operated into the 1920s and was one of the largest planemakers in the country, making many of its products for private brands. It is said to have essentially shut down after a series of unprofitable years and a destructive tornado.

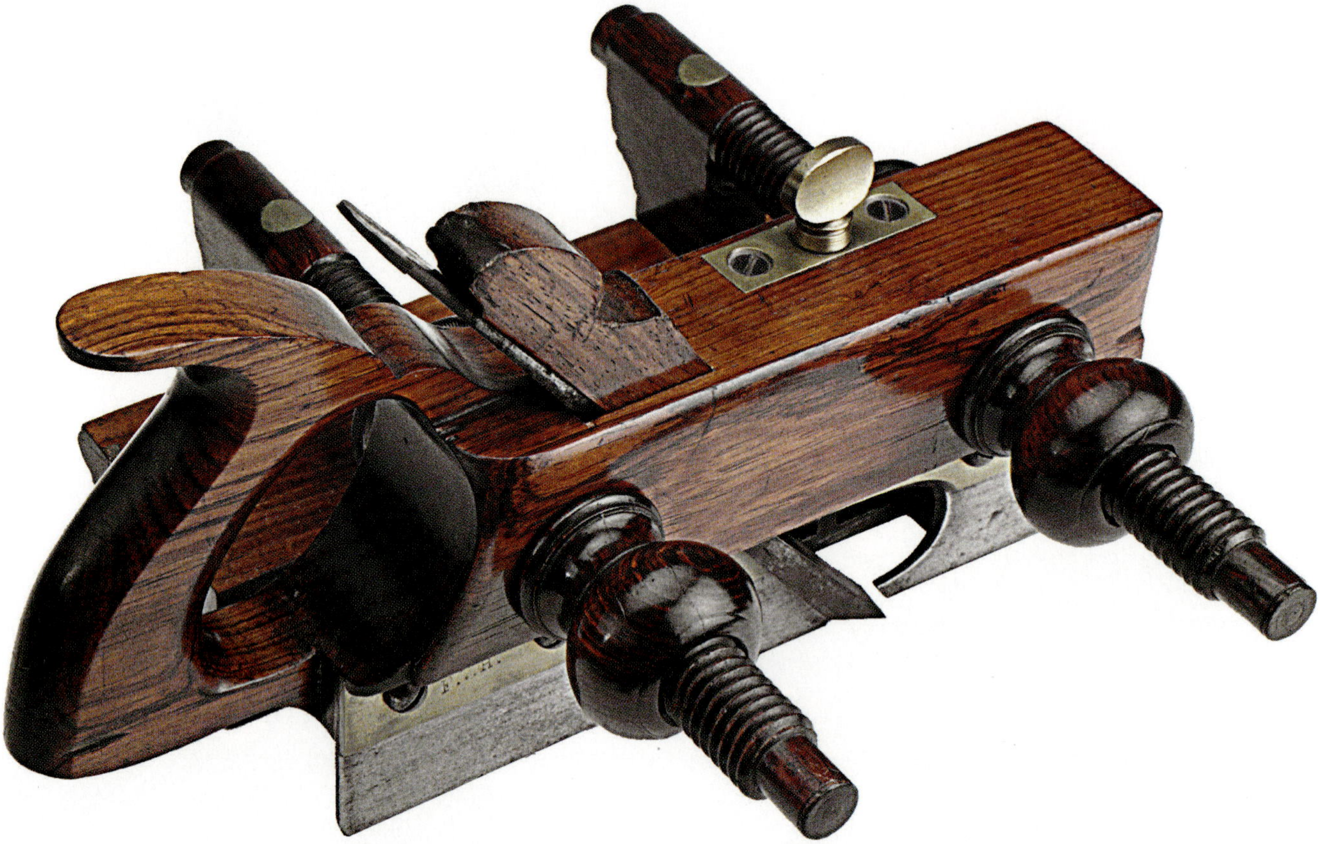

Sandusky Tool Co.

Handled rosewood screw-arm plow with fence screws going into t-nuts, and depth stop lock. Model 132

(From the collection of Kenneth Thomas <<< Don Rosebrook)

While the flat nuts remained the same, this plane carries large nuts that are substantially different from those on the preceding plane.

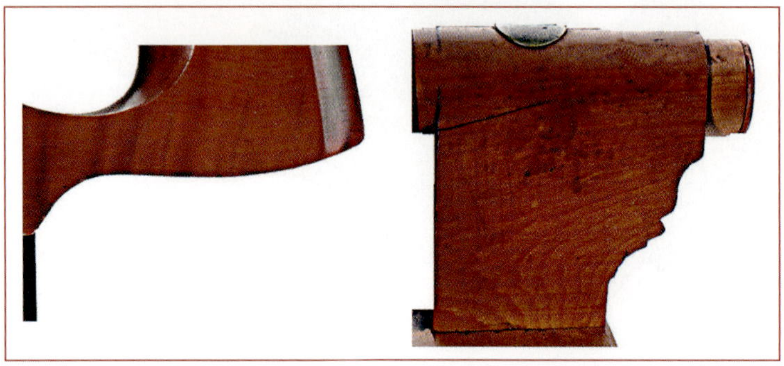

Sandusky Tool Co.

Handled boxwood screw-arm plow with fence screws going into t-nuts, and depth stop lock. Model 132

(From the collection of Don Rosebrook)

Figures or patterns in the wood, as seen on this plane, can make a relatively ordinary plane very attractive. This type of pattern is often called tiger striping.

Sandusky Tool Co.

Handled boxwood screw-arm plow with ivory tips and an ivory wear strip on the fence, and depth stop lock. Plaque on top of body reads "Ohio Sandusky Tool Co. 1872".

(From the collection of John Walkowiak)

Note that the large nuts differ from those on other Sandusky planes by having knurling on the inner rim.

This plane was entered in the Philadelphia Exposition of 1876. Below is the medal awarded to Sandusky for the plane.

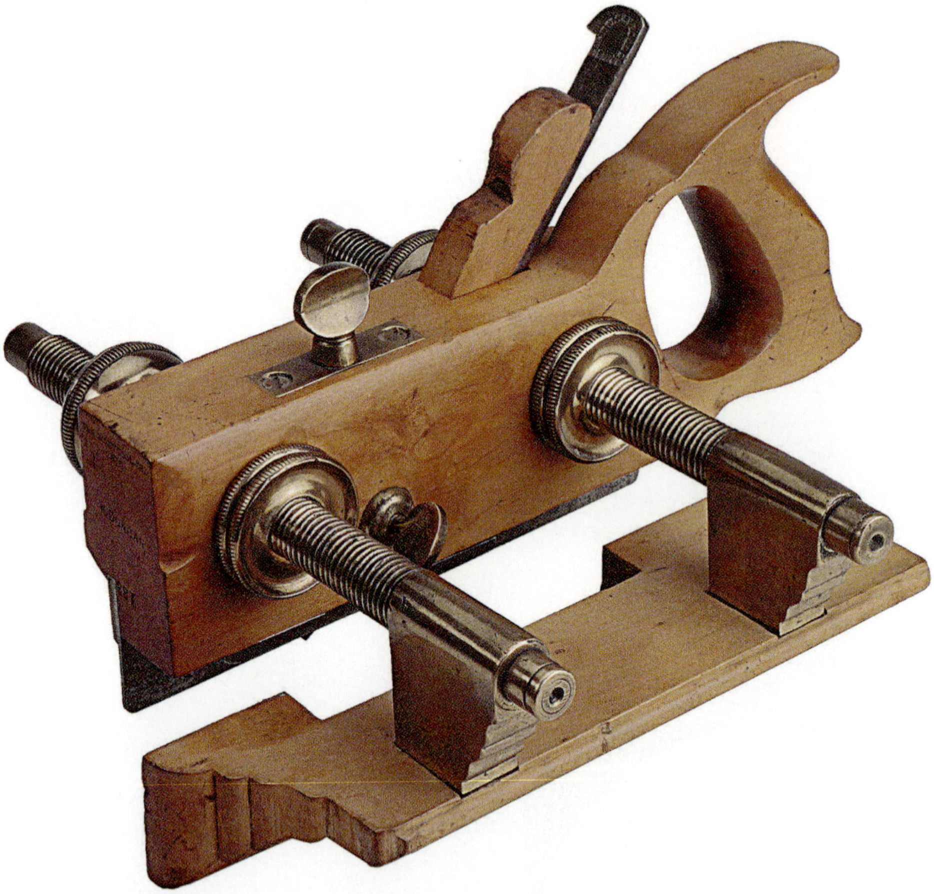

Sandusky Tool Co.

Handled boxwood screw-arm plow with brass arms and nuts. Model 133.

(From the collection of John Walkowiak)

Note the exact replication, in brass, of the shape of wooden nuts and of arms at the shoulders and tips. It appears that this is the only planemaker to incorporate shaped handles as shown on this plane.

Taylor

Unhandled beech screw-arm plow.

(From the collection of William Curtis)

Hiram Taylor was listed as a planemaker in the Cincinnati, Ohio city directories of 1842-1843 and as a planemaker's storekeeper in 1844. He was listed again as a planemaker in 1860-1861, 1864, and 1867. This probably means that he was continuously engaged as a planemaker from 1842 to 1867, and on his own from 1860-1867.

W. L. Epperson

Handled rosewood screw-arm plow with ivory tips, and silver mounting. Additional items of silvered trim include ornamentation and plaque on side. ****

(From the collection of Steven Johnson)

William Epperson was a planemaker in Louisville, Ky and according to the city directories he worked at this profession from 1858-1859, 1879 and 1890. In other years he was listed as carpenter, machinist, cabinetmaker, fishing rod maker and violin maker.

The items of ornamentation on this plane include the heart on the handle, six-pointed star on the top of the body, masonic symbol on the right side of the body, bands on the nuts, fasteners for the ivory tips, depth stop escutcheon, and threaded inserts on top of the arms. Note especially the ornamental shape under the handle. This was most probably a presentation piece.

A. McBride

Handled rosewood screw-arm plow with ivory tips, depth stop lock, and silver mounting. Additional items of silvered trim and of brass trim are present. **

(From the collection of Don Rosebrook)

Alexander McBride made planes in Louisville, Kentucky from 1851-1852. He was also listed as an architect, carpenter and, from 1851-1866, sole proprietor of a hardware business.

The items of ornamentation include the seven-pointed star on the top of the body, the threaded inserts on the top of the arms, and the five teardrop-shaped pieces opposite the screws in the skate. This plane would appear to have been a presentation piece or at least a special order plane.

5 Craftsman-Made Planes

At various times and in various places, craftsmen and skilled amateurs had need for a plow plane, and either did not have the money or the opportunity to buy a manufactured piece. Many of these men made themselves a plane or perhaps had it made by a friend. Some craftsmen may have been intrigued by a particular new development for plow planes and felt that they could produce such a tool, or that they could create a better tool.

The earliest plows (Yankee plows) were not complex. Their construction did not call for fancy moldings or fancy turned nuts, or any threaded piece. If the maker could chop mortises through the body for the arms and mortises in a featureless fence to hold the arms straight, use crude rivets to hold the arms, and had a piece of iron for the skate, then he could make a plow plane. Indeed, many of the craftsman-made planes of the 1700s and early 1800s were wedge-lock slide-arm planes. And while the maker needed only to make a functional tool to please himself, many began to add the kinds of features that graced manufactured planes by the early 1800s.

Craftsmen made screw-lock slide-arm planes with screw-lock depth stops. Moldings were added to an extended front on the fence and to the shoulders of the arms. These moldings showed varying degrees of skill as did the other features of the planes. (This was true of manufactured planes also, and was also probably a function of the skill of the teacher and the length of the apprenticeship.)

Craftsman-made planes are also known with screw-arms and constructed of exotic woods. These planes required more skill and access to more tools and better, more expensive materials. With the arrival of some of the patented planes, such as those of Israel White, some craftsmen took up the challenge to copy these innovations.

This section begins with a look at the planes of a modern planemaker, Robert Baker, a craftsman by any standard and most certainly a planemaker. Examples of his craftsmanship are also evident elsewhere, such as the restored rosewood Ohio Tool Co. plow plane.

Next, this section exhibits a group of planes that are unsigned and for which there is no obvious maker. Rather than guess at possible makers, these planes, many of which are quite interesting, are grouped together under Craftsman-made (and some may be).

The concluding pages contain some examples of earlier craftsman-made planes. These planes exhibit considerable originality and skill on the part of their makers. Some of the features represent original thinking. Some of the features represent an attempt to decorate their tools. Some of the planes represent attempts to duplicate the innovative features of manufactured tools and succeed quite well. These planes are a tribute to the early American craftsman and his talents.

The Planes of Robert Baker

Robert Baker (Bob) was born in Michigan in 1954. He is known to antique tool collectors as a restorer of antiques. He is a craftsman who strives to perform authentic restorations and encourages his clients to maintain the original integrity of those pieces that he restores. He is concerned about the accuracy of the restoration and of maintaining the original surfaces with their patinas insofar as possible.

Originally a motorcycle mechanic and racer, Bob later attended Michigan State University, graduating in Museum Studies. He began making planes in 1979 with the production of some dovetailed miter planes. In 1981, he began making reproductions of the Falconer coachmakers' plow which was, at the time, on display in the Science Museum in London. Between 1981 and 1985, he made five of these planes.

From 1983 through 1990 he produced the reproductions of the Sandusky brass center-wheel plows shown here. There are 22 of these planes. The boxwood plane with ivory tips, shown here, is one of two that he made of that wood. The pictures on the following pages show various steps in the making of the boxwood plane. The turner at the lathe is his son, Aaron, a skilled worker in his own right.

The boxwood plane was a special order tool for Andrew Delans, shown in one of the photographs, beaming into the camera upon acceptance of the plane from Bob. In addition to the boxwood models, he made five planes of ebony with ivory tips. It was only after these ebony planes were made that one, and more recently another, authentic ebony Sandusky center-wheel plane was discovered. The other planes in the set have been rosewood, and a combination of rosewood and boxwood. One of the planes was made to be a left- handed plane (i.e., a plane for a left-handed user.)

In 1985, he was fortunate enough to have Solon Rust's planemaking tools in his own shop for nine months, on loan from Ken Roberts. Using the tools and patterns from the tool chest, he produced 10 applewood planes of the third model Chapin-Rust patent type. In addition, he made two such planes in beech.

In 1997, he moved to York, Maine and set up shop again. In 2000 he began work on three Israel White style planes, using a spectacular piece of West Indian satinwood that he had obtained for that purpose in 1985. As always, he proceeded slowly and carefully, hand-crafting every piece, as well as doing his own carving and engraving. The planes were finished in late 2001 as were three more planes less ornamented and constructed of beech.

Making A Boxwood Reproduction of A Sandusky Center Wheel Plow Plane

Body blanks cut from the pattern; mortises laid out and chopped.

Holes drilled for arms.

Outer arm holes threaded for bushings. Center hole threaded for center arm.

Body moldings cut; tote cut and filed to shape.

Skate made from patterns and fitted into molded groove in the body.

Threaded bushings are turned and threaded on a metal lathe.

Arms are roughed out and then turned on a Barnes pedal powered metal lathe. Center-arm is threaded on the same lathe.

Brass center wheel is turned using hand-held and slide-rest tools on the metal lathe.

Center-wheel edges are knurled with self-made radius knurl.

Aaron Baker making nuts for arm to fence fasteners.

Fence is made, and parts are fitted up and inletted. Bridge and center-arm are fitted before arms are let into fence.

Six ivory tips are sawn from old cue balls. They are then drilled and turned on the lathe.

Plane is disassembled, sanded, and finished with varnish.

Owner (Andy Delans on the left) and maker (Bob Baker).

Robert Baker

Handled boxwood self-regulating plow with depth stop lock and ivory tips.

(From the collection of John Walkowiak<<< Andrew Delans)

Robert Baker made planes in Michigan from 1980-1997 and in York, Maine 1998-2001. This plane was made in 1989, using measurements taken from an authentic Sandusky brass center-wheel model plane. The nose of the plane is clearly marked with Robert Baker's name and the date it was produced.

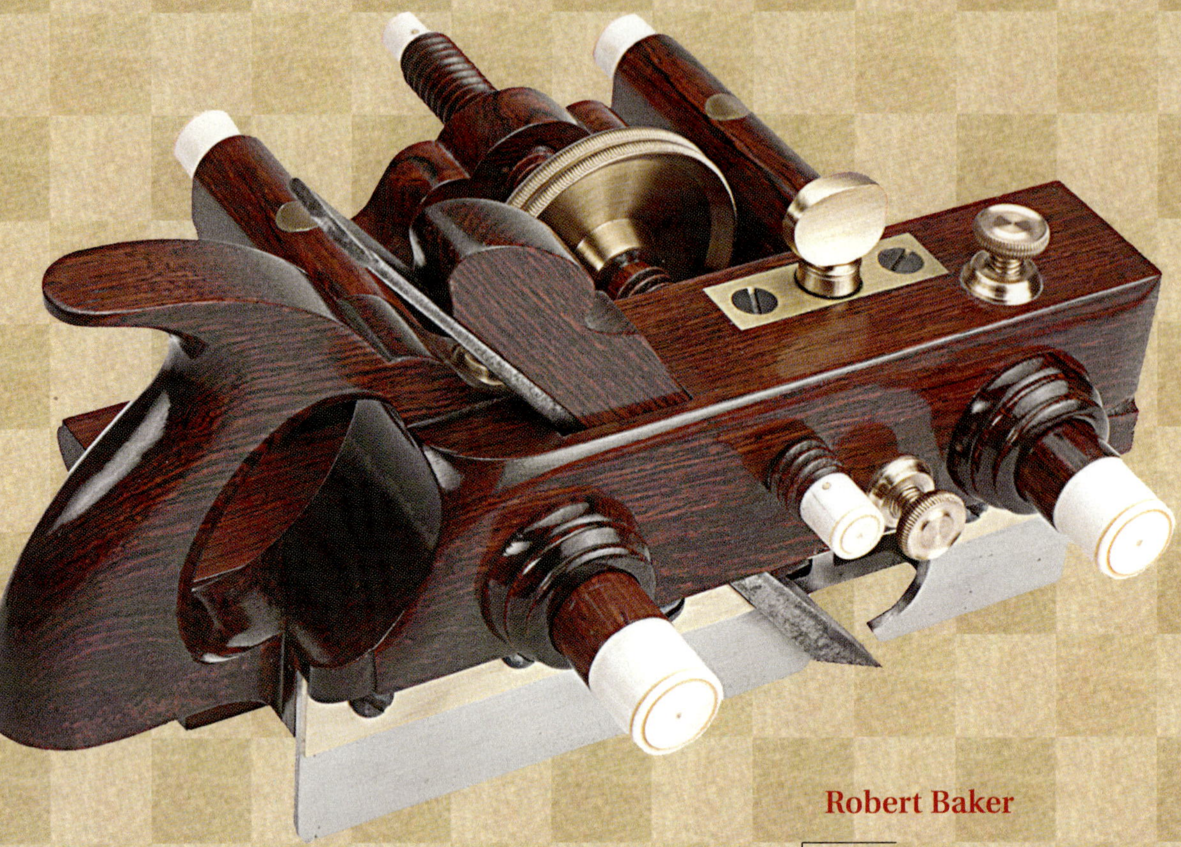

Robert Baker

Handled rosewood self-regulating plow with depth stop lock and ivory tips.

(From the collection of John Walkowiak<<< Andrew Delans)

This plane was made in 1985, using measurements taken from an authentic Sandusky brass center-wheel model plane. The nose of the plane is clearly marked with Robert Baker's name and the date it was produced.

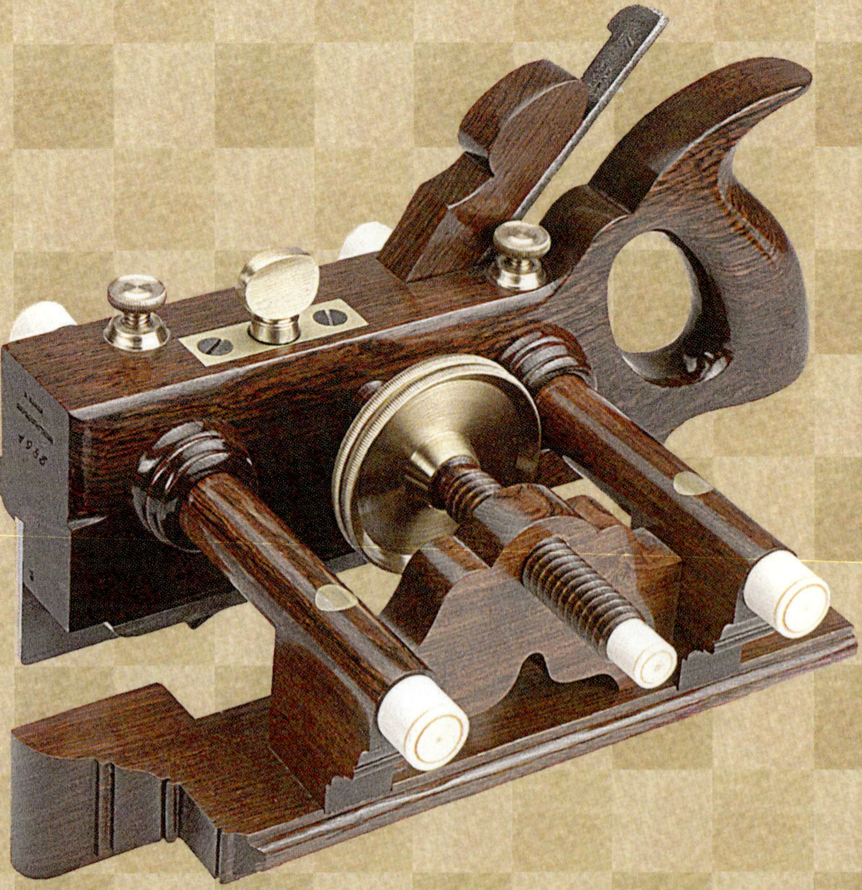

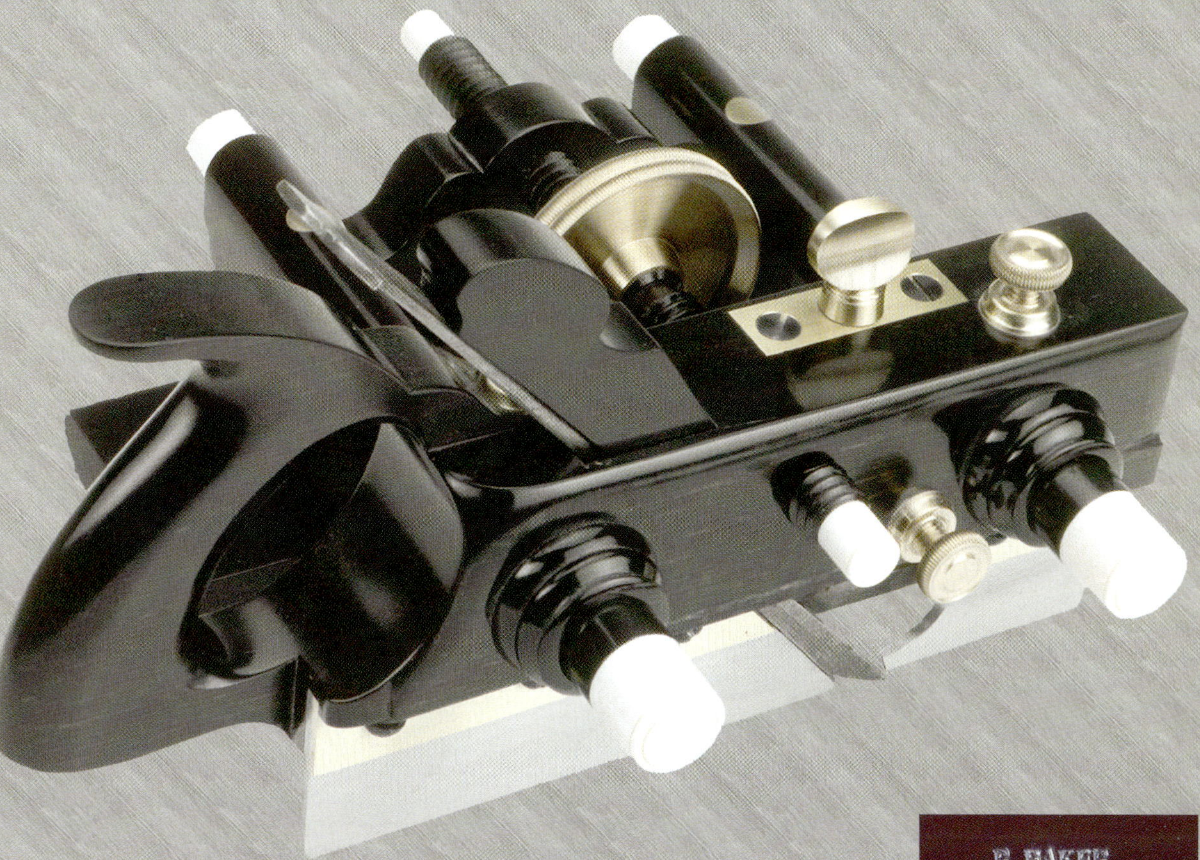

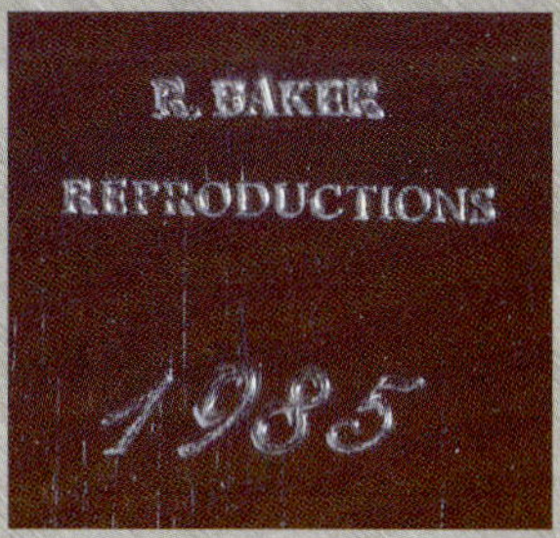

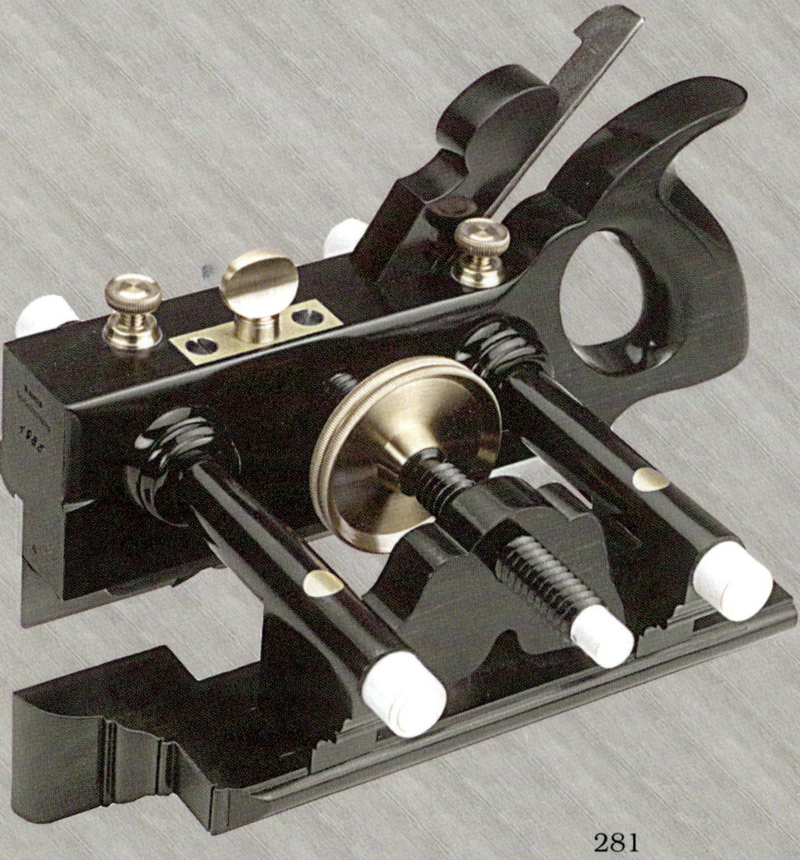

Robert Baker

Handled ebony self-regulating plow with depth stop lock and ivory tips.

(From the collection of John Walkowiak<<< Andrew Delans)

This plane was made in 1985, using measurements taken from an authentic Sandusky brass center-wheel model plane. The nose of the plane is clearly marked with Robert Baker's name and the date it was produced.

Robert Baker

Unhandled Israel White style West Indian satinwood plow with ebony arms and lignum vitae boxing on the fence, ivory tips and scales, ivory eagle, and sterling silver trim. One of three.

(From the collection of Don Rosebrook)
Photographs this page and next by Larry Weaver.

There was only enough of this spectacular West Indian satinwood to make three planes. Every piece of the plane, including the screws, was crafted by Mr. Baker. Silver pieces are solid silver, not silver plate. The eagle and the border around the scale for the depth stop, are carved ivory, which was then inlaid into the plane. All engraving was done by the maker, who clearly marked the nose of the plane as his product.

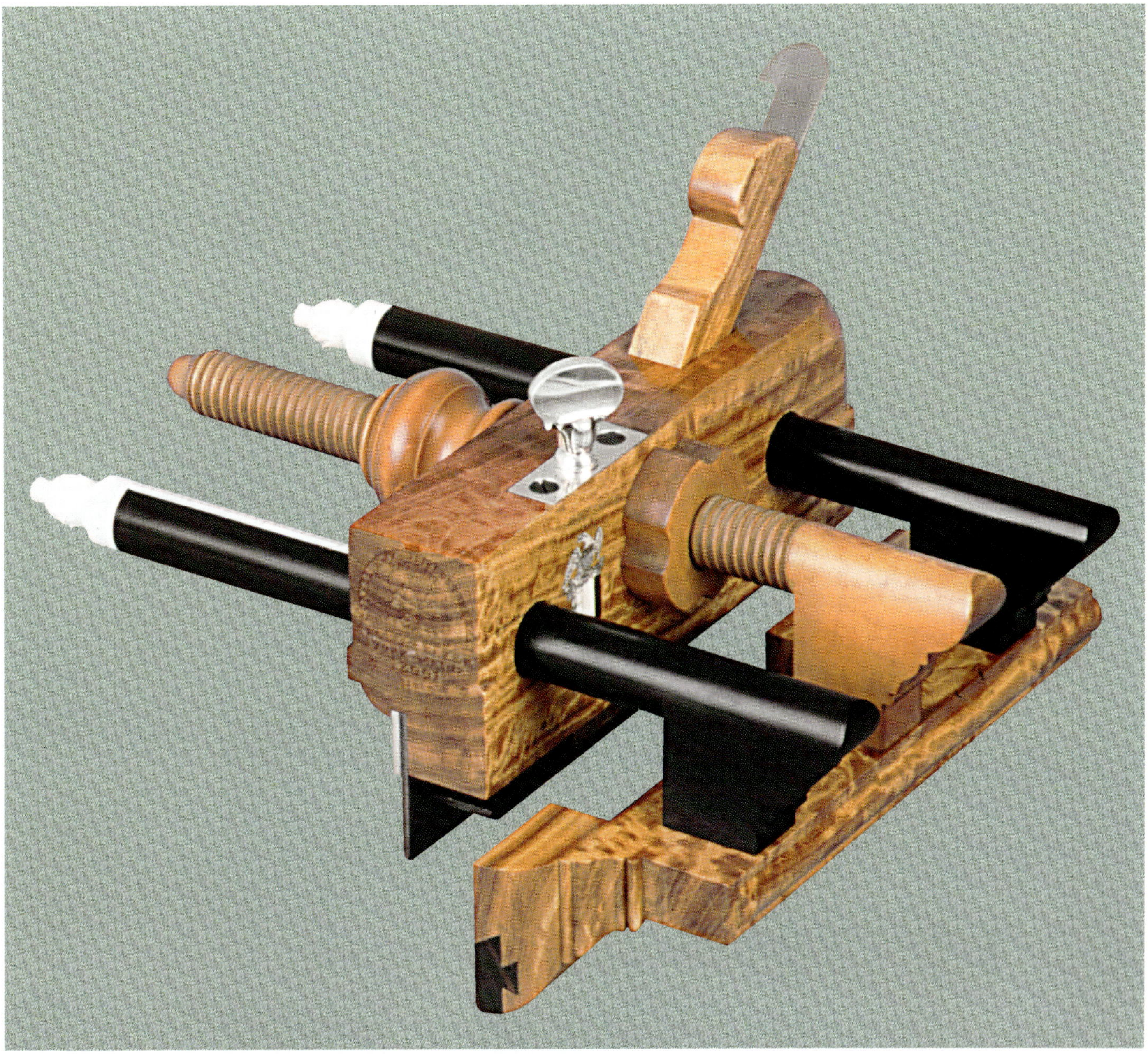

WHITE
ROBERT BAKER
YORK MAINE
2001

Unsigned

Handled boxwood screw-arm plow with ebony nuts and silver mounting. Knurled thumb screw lock under depth stop adjustment screw.

(From the collection of Don Rosebrook)

This is a remarkably attractive plane because of the ebony trim, silver mounting and the extra detail such as the molding cut under and around the base of the handle.

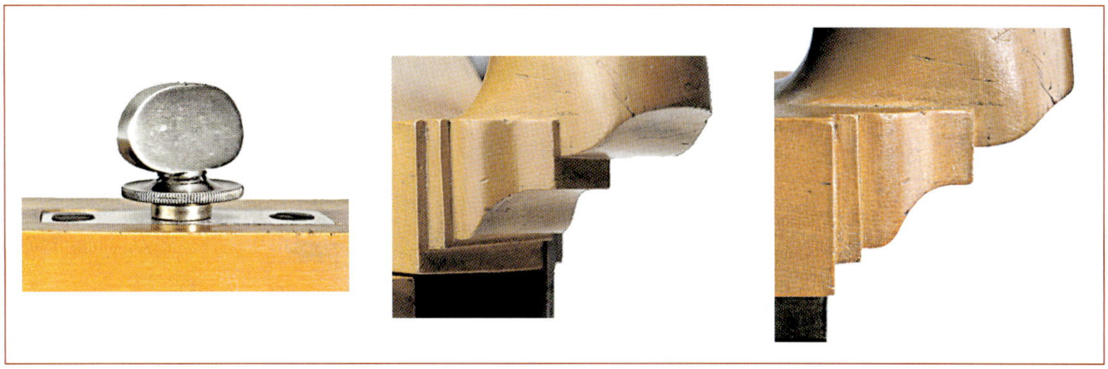

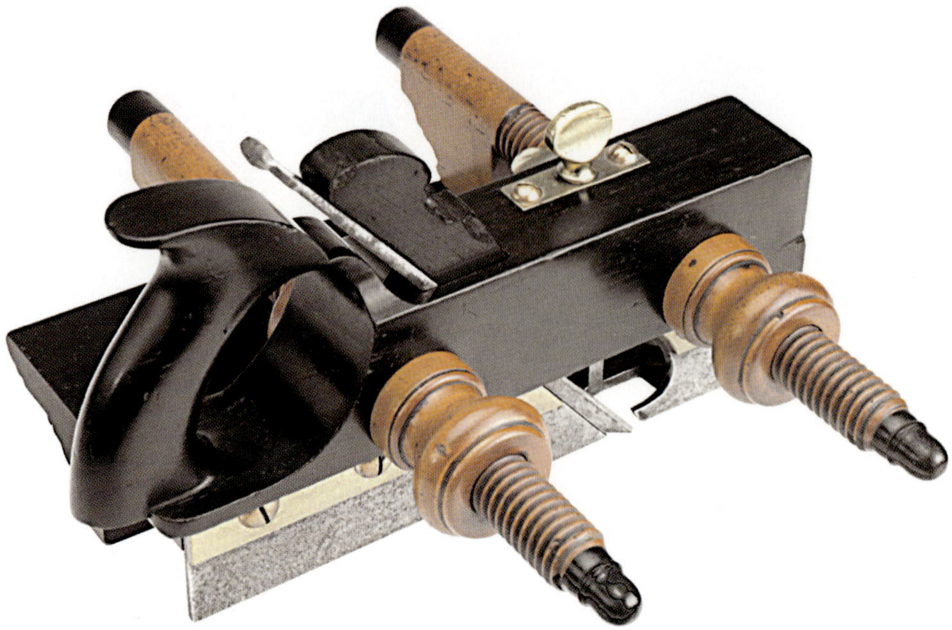

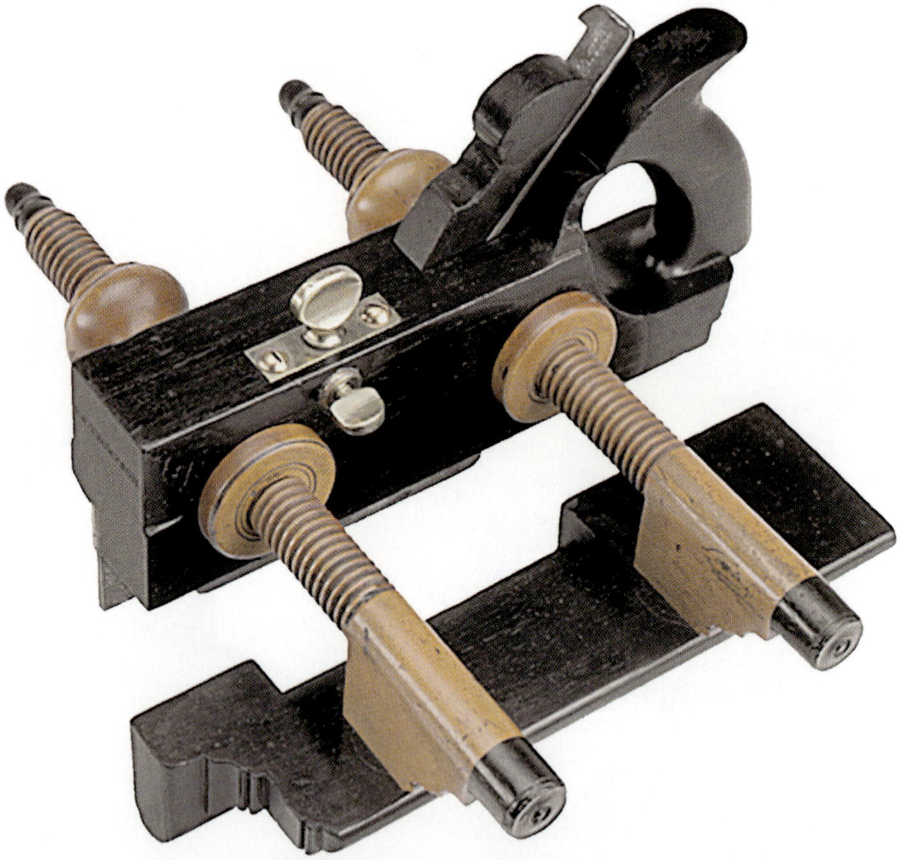

Unsigned

Handled ebony screw-arm plow with boxwood arms and nuts, ebony tips and silver mounting.

(From the collection of Don Rosebrook)

This is an owner modified plane. The body appears to be from a DeForest plane, while the fence and arms resemble those made by Denison and the ebony tips appear owner made.

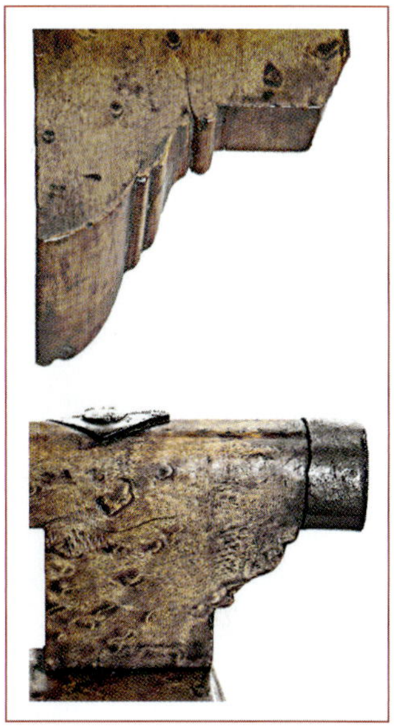

Unsigned

Unhandled birds-eye maple wedge-lock, slide-arm plow with iron collars on the arms.

(From the collection of Don Rosebrook)

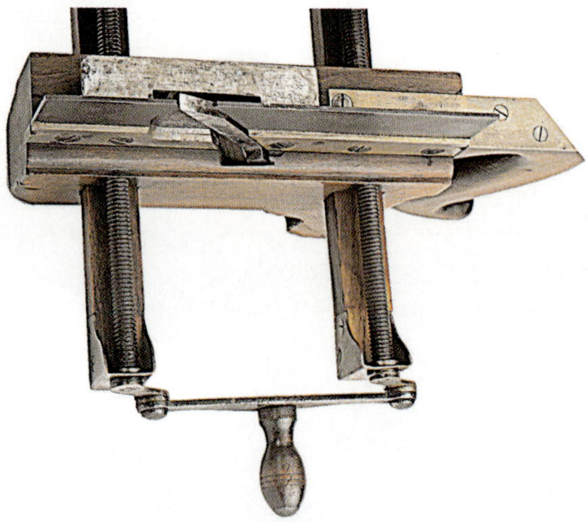

Unsigned

Handled beech plow with oak wedge and crank adjusted fence. Handle is separate piece connected to body with brass plates.

(From the collection of Durward Gehring)

A look at the bottom of the plane shows the mechanics of the system. However, like most cranks with this design, it will jam when it is turned rapidly. There is no provision for locking the fence after it has been adjusted.

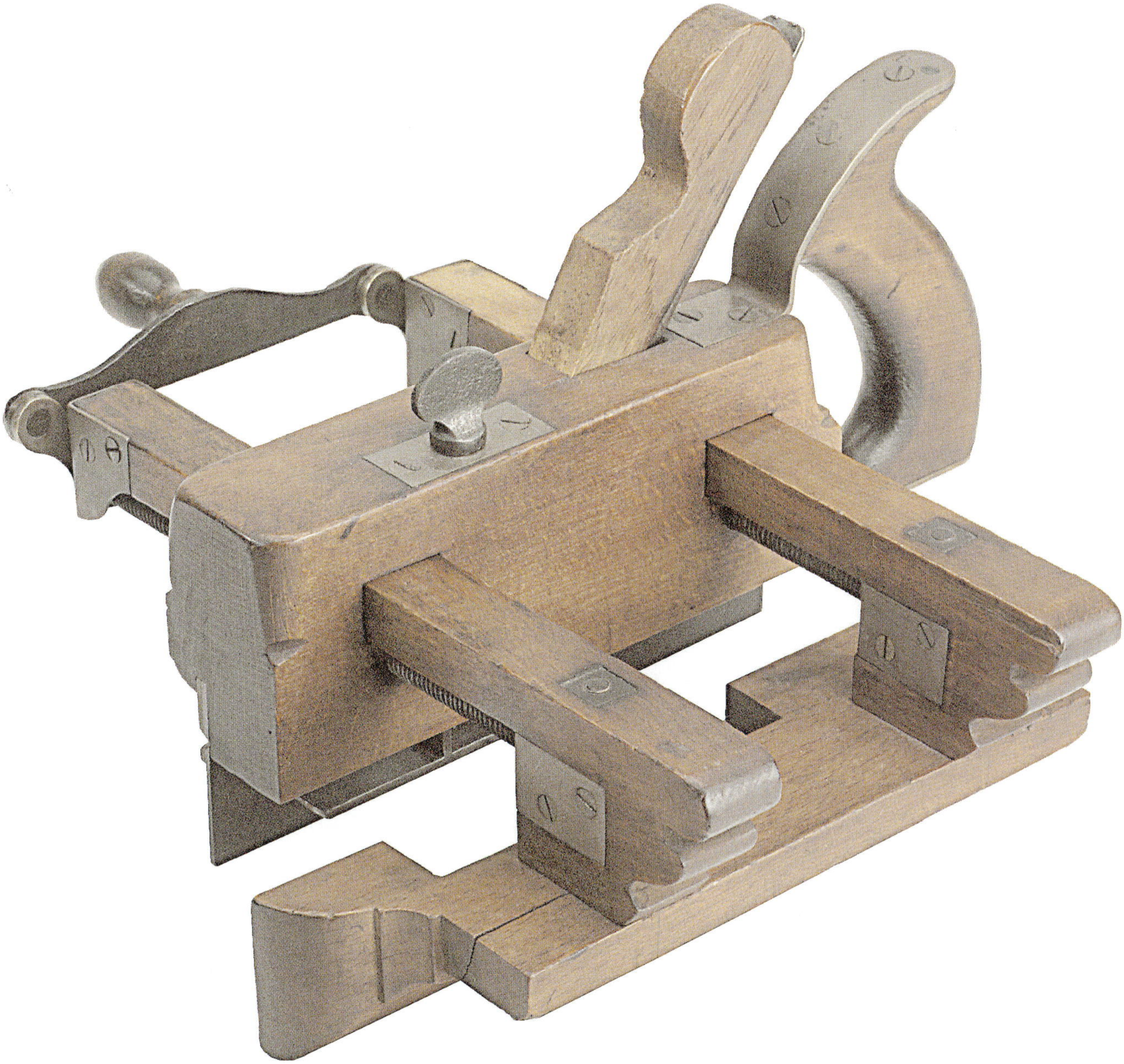

Rather than just being an unsigned experimental plane or prototype, this may well be a craftsman made plane.

Unsigned

Handled boxwood screw-arm plow with rosewood fence and wedge and oversize ivory tips.

(From the collection of Peter Hathaway)

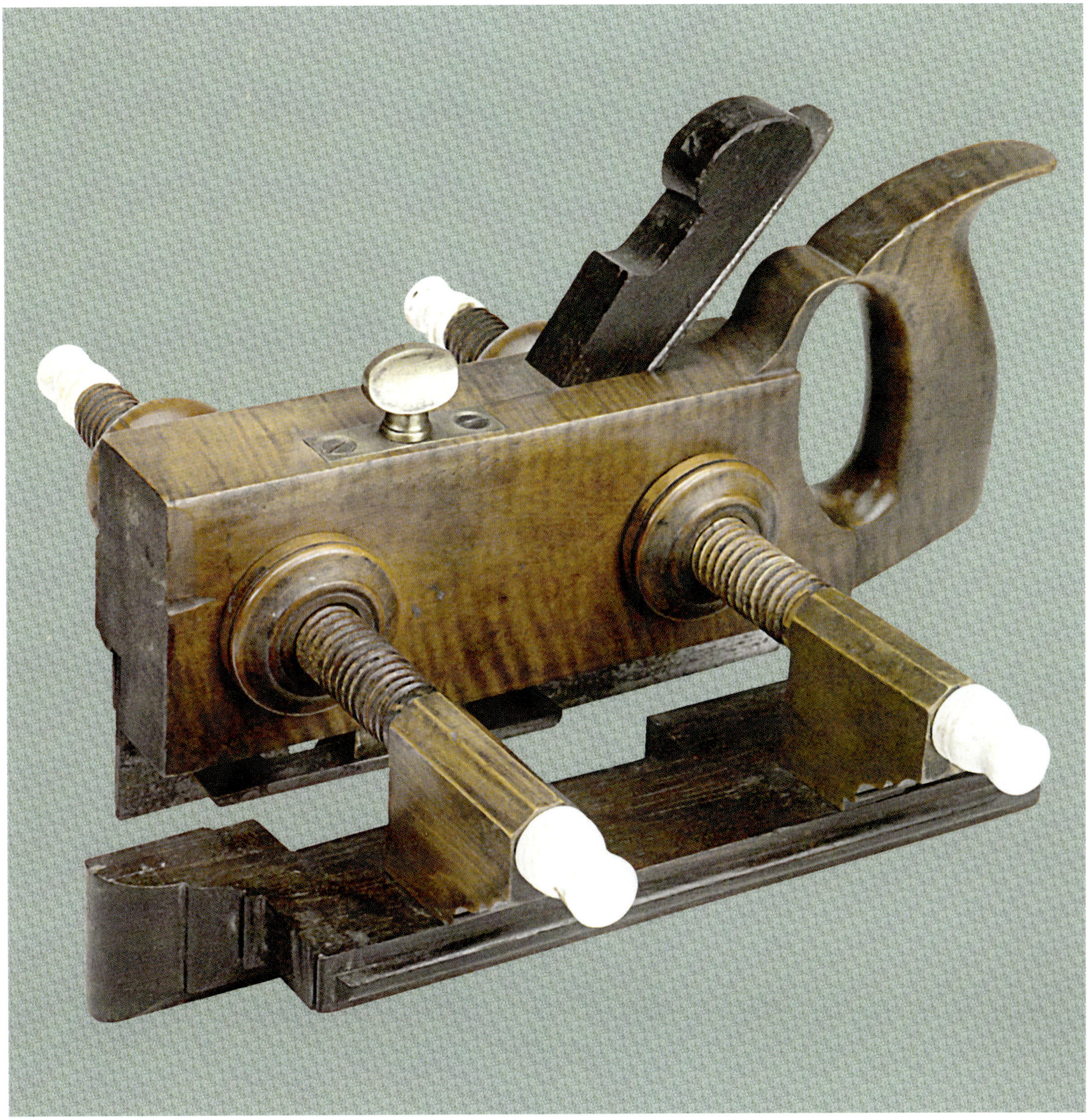

The oversized ivory tips look strangely out of place on this plane and may have been owner added, but the remainder of the plane appears to have been made by a manufacturer.

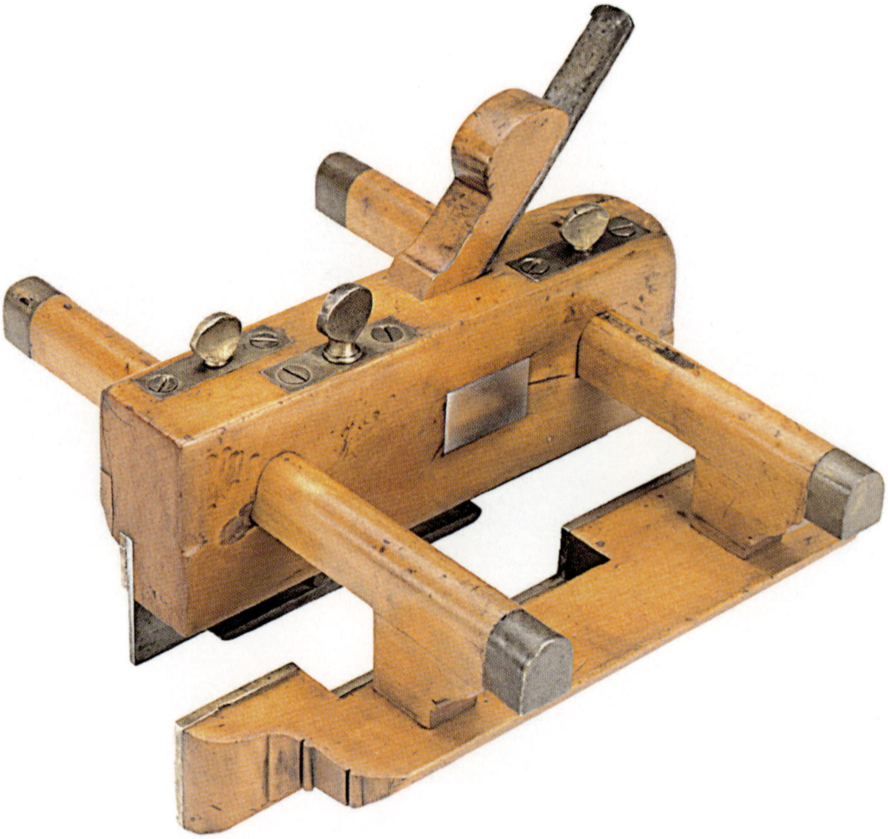

Unsigned

Unhandled boxwood screw-lock slide-arm plow with unusual locking system and blank plaque inlaid on side. Brass faced fence and brass caps on the arms.

(From the collection of Steve Dice)

The authors have not seen another plane with this type of locking system. The effect is to apply pressure to the bottom of the arm, rather than the top like a screw-lock or to the side like a wedge lock. Judging from the plaque in the side, this plane was meant to be a presentation piece.

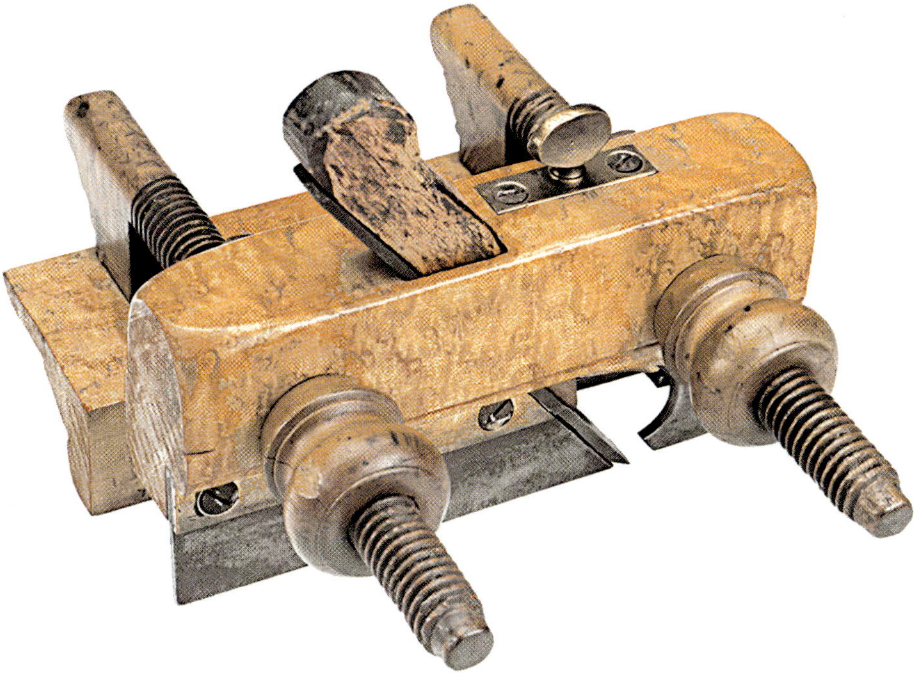

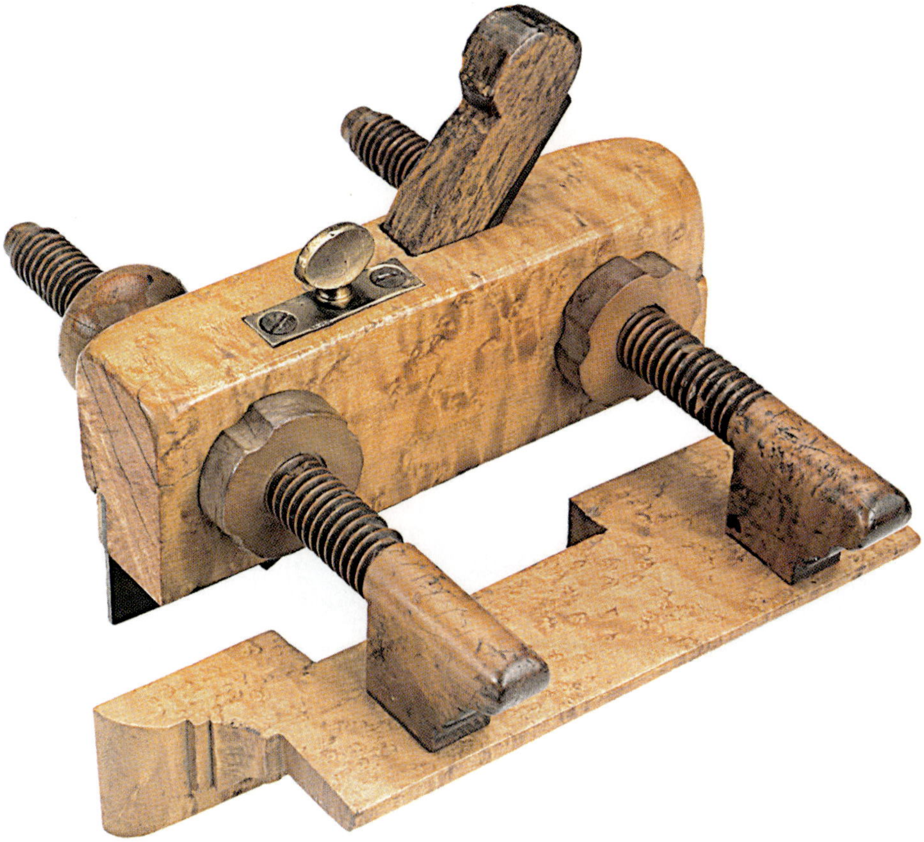

Unsigned

Unhandled birdseye maple screw-arm plow with boxwood arms and nuts.

(From the collection of Steve Dice)

The wedge is probably not original to the plane.

Craftsman-Made

Unhandled fruitwood screw-lock slide-arm plow with threaded pewter inserts for thumbscrews, and bone inlay.

(From the collection of Don Rosebrook)

The use of bone for decoration while not unknown was certainly uncommon. In this case, bone was also used for a wear strip to protect the top of the arm. The use of threaded inserts made of pewter is unique in the authors' experience. With features such as square arms mortised into the fence, the length and simplicity of the fence, and the shape of the wedge, this would appear to be a very early plane. However, it does not have the flat chamfers that are usually characteristic of the early Yankee plows.

Craftsman-Made

Handled maple screw-arm plow.

(From the collection of Steve Dice)

The body of this plane, exclusive of the handle, is 10 3/4" long, making it 3" or 4" longer than the standard manufactured handled plane. The use of maple is not unusual in that craftsman planemakers used whatever woods were conveniently available to them.

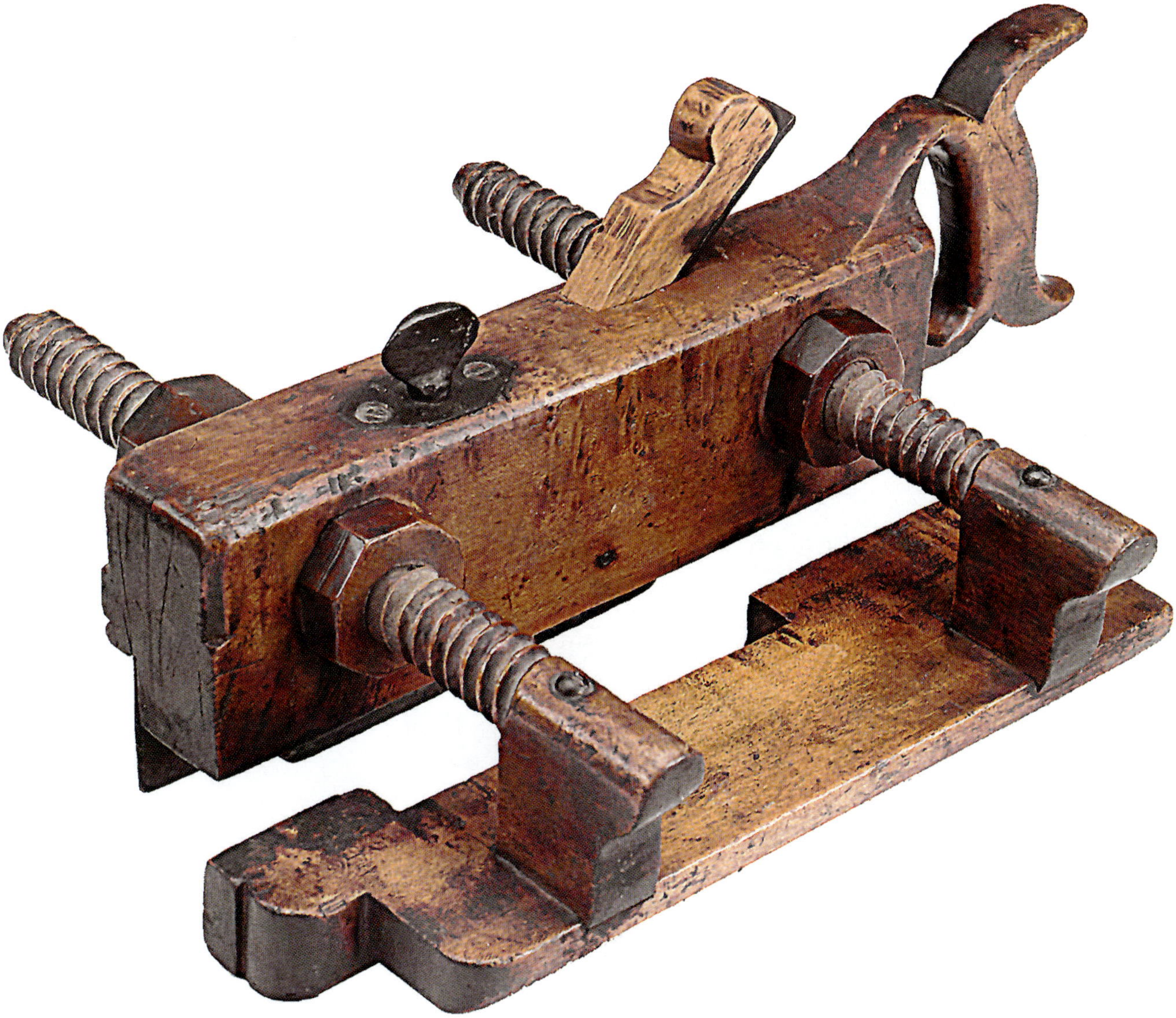

The use of hex nuts and the coarseness of the threads are also features that mark this as a craftsman-made plane. Close observation indicates that the threads (internal in the nuts and external on the arms) were meticulously hand-cut by the maker. The handle is unusual, to say the least, and the depth stop is hand wrought.

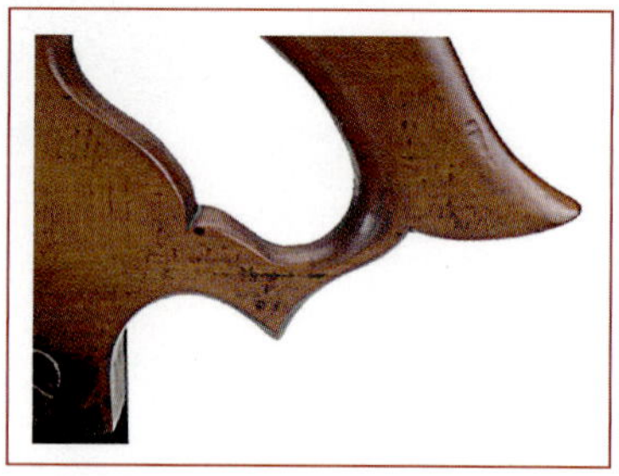

Craftsman-Made

Handled maple slide-arm plow.

(From the collection of Steve Dice)

This plane has no apparent way to lock the arms, and thus the fence, in place. However, there are holes in the body near each arm and holes in each end of the arms. It would seem that there once have been some type of mechanism that could be used to lock the arms.

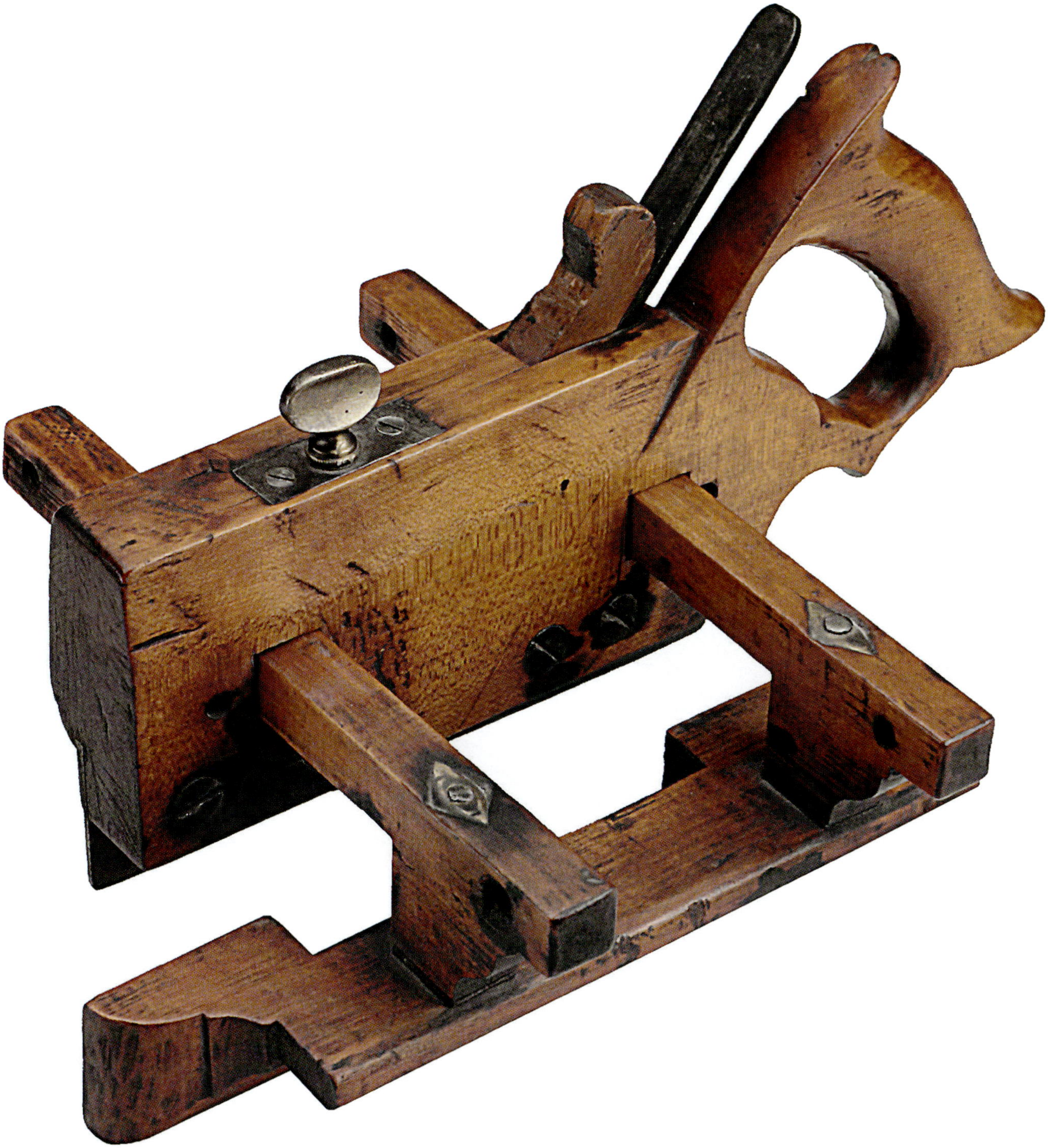

Here again, the handle is unusual. Another unusual feature is that the screws that hold the skate originate on the right side of the plane. This means that the holes in the skate must be threaded or that the ends of the screws are battered down to act as rivets.

Craftsman-Made

Unhandled lignum vitae screw-lock slide arm plow.

(From the collection of Steve Dice)

The use of lignum vitae in planemaking is quite uncommon because the wood is difficult to work with hand tools and is prone to subsequent checking. Another unusual feature is the use of round arms. The thumbs screws are crude and obviously hand made.

Craftsman-Made

Unhandled rosewood screw-arm plow with brass bottom.

(From the collection of Frank Manville)

Although technically this plane differs from the other plow planes in this book, it is attractive and very interesting. It is used in shipbuilding and is called a ship hawk. The plane can only be used to plow a V groove. The threads on the arms are as fine as one would expect to find on metal arms. The use of a brass piece on the bottom is very unusual as is the design of the depth stop, which is external to the body. Finally, the very open center of the body in the front of the wedge is unusual for any wooden plane.

Craftsman-Made

Handled fruitwood screw-arm plow, unplated, with brass wear strip on the fence.

(From the collection of Peter Hathaway)

Beginning with the simplistic fluted shape of the large nuts, this plane shows the skill of the craftsman who made it. The overall shape of the body and handle is generally similar to that of manufactured plows, but the fence and the arms are particularly well done. Note that there is a wing nut on the depth stop adjustment rather than a thumb screw.

Craftsman-Made

Handled beech slide-arm plow with brass arms.

(From the collection of John Sindelar)

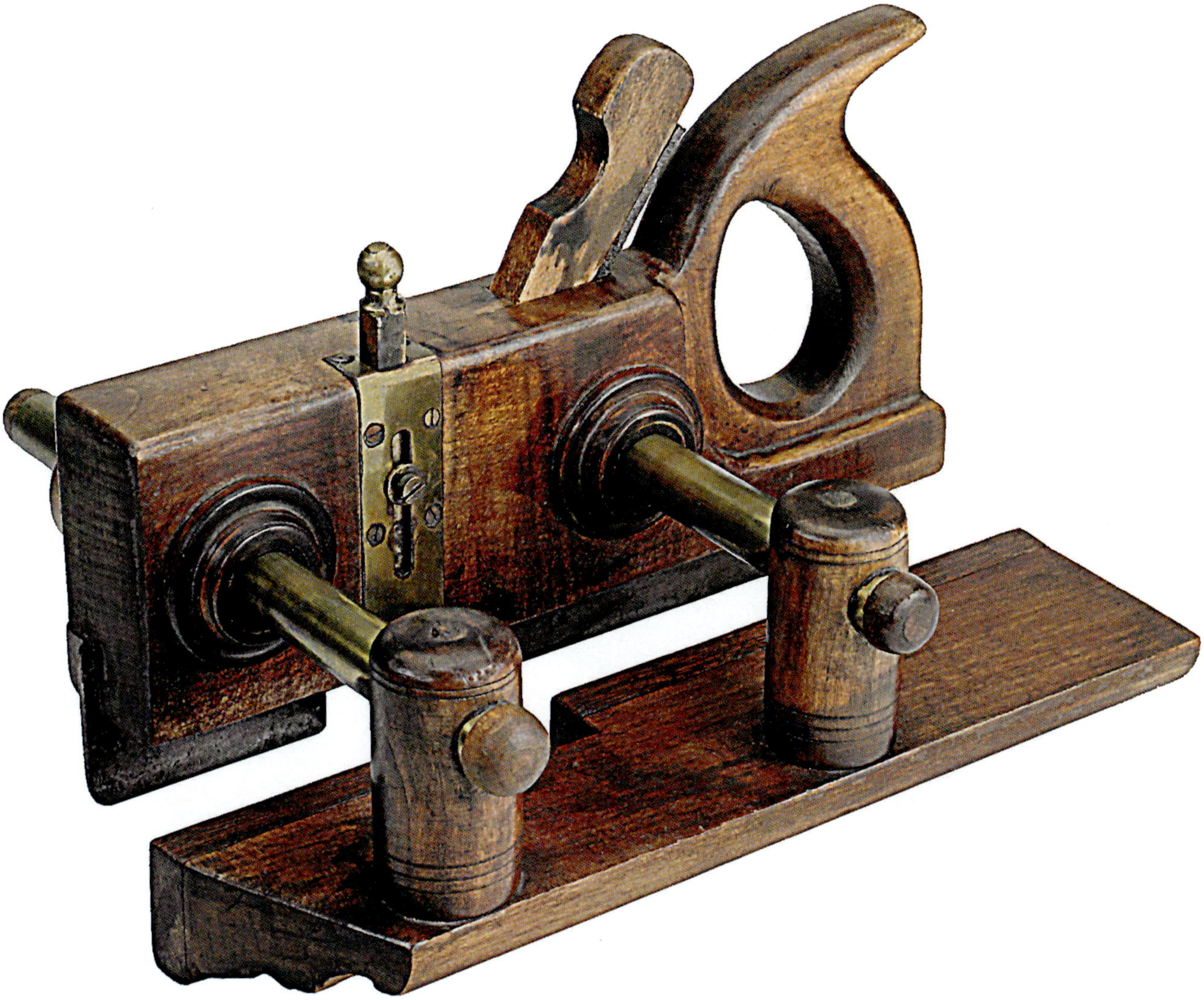

This plane has a unique construction. The arms are tubular brass and the nuts are bushings (sleeves). The outermost nuts contain set screws to lock the fence in place. The fence is of two piece construction and appears as if a wear strip was attached to the edge of a piece of crown molding. The depth stop, shown on the facing page, slides down the outside of the body. Round arm posts (arm shoulders) present a pleasing complement to the round arms.

Craftsman-Made

Handled beech self-regulating plane. Owner's mark is A. L. Soles.

(From the collection of Steve Dice)

We have recorded the owner's mark on this plane because it is so well made. The adjustment mechanism appears to incorporate features found on the patented Israel White planes (which see). However, this is somewhat deceiving in that the adjustment knob is on the side of the body opposite from that of all the manufactured three-arm planes. The knob is captured at the surface plate just as an Israel White. It then passes through the body and into a threaded nut inside of the fence. Turning the knob will thus move the fence. Wedges are used to secure the arms. The ruled strips inlaid into the arms are boxwood. The bridle is massive, as is the body at 12" long. The long skate is plated with a steel piece. The screws appear to be hand made, which implies that the plane was made before about 1840, but not much before, because the Israel White patent (probably the inspiration for this plane), was granted in 1834.

The boxwood scales in the arms are each 3½" long and are rounded to conform to the surface of the arms and this implies that the scales were made especially for this purpose (as opposed to being a section of a boxwood folding rule).

Craftsman-Made

Unhandled beech self-regulating slide-arm plow.

(From the collection of Kendall Bassett)

This plane represents another attempt to copy an Israel White patented plow. The body and the fence seem very well done, and may have been taken from a manufactured screw-arm plow plane. The fence is faced with steel and the skate also has steel plating. The bridge and screw seem less well made and there is no provision for locking the fence after it has been adjusted.

Craftsman-Made

Handled beech hermaphroditic screw-arm/slide-arm plow with a beehive nut on the screw arm and a thumb screw lock on the slide-arm.

(From the collection of Roger Springate)

The construction of this plane is unique in the authors' experience. There does not appear to be any advantage to be gained by the use of one screw-arm and one slide-arm. The left hand end of the screw-arm serves as a side handle for the plane and it is captive in the body of the plane. The fence is threaded so that it moves when the handle is turned; it is locked in place by the nut on the outer side of the arm.

6 Galleries

The galleries are collections of photographs of large nuts on screwarm plow planes, shoulders of arms from plow planes, examples of different shaped skate noses, and examples of boxing.

This work is an extension of the pioneering effort begun by John Moody in 1981 wherein he presented tracings of the profiles of 41 large nuts from different planes.[1] He likewise presented tracings of the fence molding profiles of 69 different planes. In the following galleries, we have presented photographs of 93 different nuts and 83 different shoulder molding profiles. In addition, there are twelve different skate noses and six different types of fence boxing.

The shape and style of the large nuts on planes are perhaps the most distinctive characteristic of a given maker's design. The fence profiles and the arm shoulder profiles are also distinctive but not to the same extent. Some of the profiles are so outstanding that the maker of the plane can be identified from across the room if the profile is silhouetted. The authors have found that by using the information presented here, i.e., nuts, shoulder profiles, and skate noses, a near-certain assignment of identity can be made to any of the makers included.

The photos of the large nuts include the arm tip and this information is also valuable. Additional information in the form of the style of the flat nuts, the fence molding and the under-handle shape (for many planes) may be found on the pages dedicated to a specific plane.

If any more reason is needed for presenting the galleries, the beauty and grace of the nuts and arm shoulders should be sufficient incentive. The artistry of the early American planemakers is obvious.

[1] Moody, John A. "*The American Cabinetmaker's Plow Plane.*" The Tool Box (1981).

Arrowmammett

L. B. Bigelow

H. Chapin

H. Chapin

H. Chapin

DeForest

DeForest

DeForest

G. W. Denison

J. Denison

J. Denison

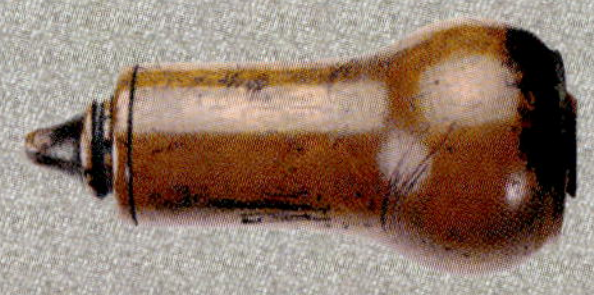
J. Denison

C. Morehouse

Shelton & Osborn

Shelton & Osborn

Shelton & Osborn

Shelton & Osborn

Winstead Plane Co.

Epperson

McBride

H. L. Kendall

R. W. MacCubbin

Thos. Appleton

M. Copeland

A. Cumings

Gladwin & Appleton

P. A. Gladwin

Greenfield

Greenfield

Greenfield

Greenfield

Greenfield

Greenfield

Greenfield

H. L. James

James Kellogg

J. H. Lamb

J. H. Lamb

Multiform

Parker & Hubbard

Taber Plane Co.

Taber Plane Co.

Wing Taber

D. P. Sanborn

C. Warren

C. Warren

S. Cook

Auburn Tool Co.

Auburn Tool Co.

D. R. Barton

D. R. Barton

Edward Carter

L. Case

Casey & Co.

Casey, Clark & Co.

Casey, Clark & Co.

Casey, Clark & Co.

Casey, Kitchell & Co.

Casey, Kitchell & Co.

Collins

M. Crannell

De Valcourt

Doscher

Gleason

Gregory

J. Hannan

A. Howland & Co.

A. Howland & Co.

A. Howland & Co.

A. Howland & Co.

T. J. McMaster

Z. J. McMaster

Z. J. McMaster

A. Miller

Randall & Bensen

N. Spaulding

Treman & Bros.

Ohio Tool Co.

Sandusky Tool Co.

Sandusky Tool Co.

E. W. Carpenter

D. Colton

F. Dallicker

W. Goldsmith

B. Sheneman

G. White

Israel White

Israel White

Arrowmammett

Bigelow

H. Chapin

DeForest

DeForest

DeForest

G. W. Denison

J. Denison

J. Denison

Killiam

D. Morehouse

Shelton & Osborn

Shelton & Osborn

Winstead Plane Co.

Epperson

McBride

H. L. Kendall

Thos. Appleton

M. Copeland

A. Cumings

Gladwin & Appleton

Greenfield

H. L. James

James Kellogg

Lamb & Brownell

Lamb

Multiform

Parker & Hubbard

E. Smith

Taber Plane Co.

Wing Taber

Clifford

Sanborn

Warren

Andruss

S. Cook

Alford

Auburn Tool Co.

Auburn Tool Co.

Benton Evans

E. Carter

L. Case

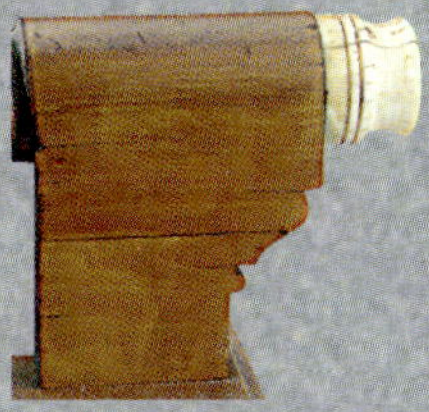
Casey & Co.

Casey, Clark & Co.

Casey, Clark & Co.

Casey, Clark & Co.

Casey, Kitchell & Co.

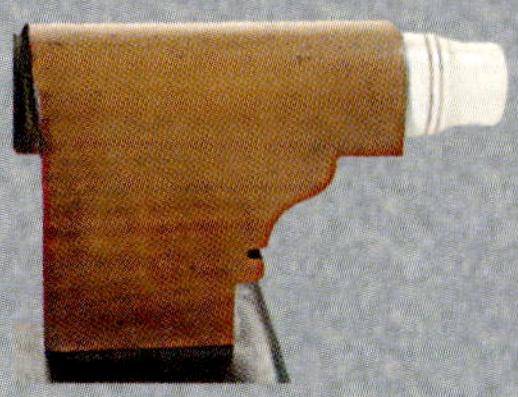
Casey, Kitchell & Co.

R. J. Collins

M. Crannell

De Valcourt

Gleason

Gregory

J. Hannan

A. Howland & Co.

A. Howland & co.

A. Howland & co.

T. J. McMaster

Z. J. McMaster

Z. J. McMaster

A. Miller

Randall & Bensen

Scovill

See

Spaulding

Treman & Bros.

Ohio Tool Co.

Sandusky Tool Co.

Seybold

H. Taylor

Bush

E. W. Carpenter

D. Colton

W. Goldsmith

Mack

B. Sheneman

T. J. Sheneman

G. White

Israel White

Battey

J. E. Child

P. H. Manchester

Pearce

I. E. Smith

Benton Evans

Chapin

J. E. Child

Doscher

Kellogg

Lamb & Brownell

Lamb

P. H. Manchester

Pearce

Taber

Thos. Appleton

T. J. Sheneman

Andruss

D. Colton

Greenfield Tool Co.

J. H. Lamb

D. P. Sanborn

B. Sheneman

Index of Planemakers